Increasing Aircraft Carrier Forward Presence

Changing the Length of the Maintenance Cycle

Roland J. Yardley, James G. Kallimani,
John F. Schank, Clifford A. Grammich

Prepared for the United States Navy

Approved for public release; distribution unlimited

RAND NATIONAL DEFENSE RESEARCH INSTITUTE

The research described in this report was prepared for the United States Navy. The research was conducted in the RAND National Defense Research Institute, a federally funded research and development center sponsored by the Office of the Secretary of Defense, the Joint Staff, the Unified Combatant Commands, the Department of the Navy, the Marine Corps, the defense agencies, and the defense Intelligence Community under Contract W74V8H-06-C-0002.

Library of Congress Cataloging-in-Publication Data

Increasing aircraft carrier forward presence : changing the length of the maintenance cycle / Roland J. Yardley ... [et al.].
p. cm.
Includes bibliographical references.
ISBN 978-0-8330-4407-5 (pbk. : alk. paper)
1. Aircraft carriers—United States—Maintenance and repair. 2. United States. Navy—Operational readiness. I. Yardley, Roland J.

V874.3.I53 2008
359.9'4835—dc22

2008008899

The RAND Corporation is a nonprofit research organization providing objective analysis and effective solutions that address the challenges facing the public and private sectors around the world. RAND's publications do not necessarily reflect the opinions of its research clients and sponsors.

Cover photo courtesy of the U.S. Navy

Published 2008 by the RAND Corporation
1776 Main Street, P.O. Box 2138, Santa Monica, CA 90407-2138
1200 South Hayes Street, Arlington, VA 22202-5050
4570 Fifth Avenue, Suite 600, Pittsburgh, PA 15213-2665
RAND URL: http://www.rand.org/
To order RAND documents or to obtain additional information, contact
Distribution Services: Telephone: (310) 451-7002;
Fax: (310) 451-6915; Email: order@rand.org

Preface

Aircraft carriers are a powerful and versatile element of U.S. naval forces. They allow the Navy to undertake a wide range of tasks, such as bringing airpower to bear against opponents, deterring adversaries, engaging friends and allies, and providing humanitarian assistance. Aircraft carriers, like other naval ships, go through a cycle of training to gain and sustain readiness, deploy to a forward theater, return from deployment, and maintain readiness to surge (i.e., to get underway to provide additional forward presence as requested by theater commanders). They also undergo scheduled maintenance at shipyards. Because carriers are among the most complex weapon systems operated by the Navy, their crews require a great deal of training and the ships demand extensive maintenance.

Depot maintenance periods consist of large and complicated work packages. The duration of maintenance periods, the type of maintenance required, and maintenance period scheduling affect the carrier fleet in numerous ways. Because personnel tempo policies have limited carriers to just one 6-month deployment per cycle, the length of that cycle affects the carrier's operational availability. While longer cycles could decrease the proportion of time a carrier is in maintenance and increase its operational availability, longer cycles with only one deployment per cycle effectively decrease the time a carrier is deployed.

In recent years, the Navy has lengthened the duration of the maintenance cycle for carriers, effectively trading actual deployment time for time that a carrier is not deployed but is able to surge. This tradeoff has made it difficult for the Navy to satisfy the combatant commanders' need for sustained carrier presence in their theaters of operation.

Recognizing this problem, the Assessments Division of the Deputy Chief of Naval Operations for Resources, Requirements, and Assessments (OPNAV N81) asked RAND to examine the feasibility and implications of increasing the forward presence of carriers by examining alternative cycles, including two deployments per cycle, and their impact on major depot maintenance work without changing deployment policies. This monograph describes the research findings. It should be of interest to Navy organizations concerned about the operations and maintenance of naval ships, especially of aircraft carriers.

The research was sponsored by OPNAV N81 and conducted within the Acquisition and Technology Policy Center of the RAND National Defense Research Institute, a federally funded research and development center sponsored by the Office of the Secretary of Defense, the Joint Staff, the Unified Combatant Commands, the Department of the Navy, the Marine Corps, the defense agencies, and the defense Intelligence Community.

For more information on RAND's Acquisition and Technology Policy Center, contact the Director, Philip Antón. He can be reached by email at atpc-director@rand.org; by phone at 310-393-0411, extension 7798; or by mail at the RAND Corporation, 1776 Main Street, Santa Monica, California 90407-2138. More information about RAND is available at www.rand.org.

Contents

Figures

Tables

Summary

The U.S. Navy currently maintains a fleet of 11 aircraft carriers. These ships, which are among the most powerful and versatile elements of U.S. naval forces, allow the Navy to undertake a wide range of tasks. They are also among the most complex weapon systems operated by the Navy. The carriers themselves need continuous and regularly scheduled maintenance. Their crews require a great deal of training to attain and sustain readiness levels. The *length* of the training, readiness, deployment, and maintenance cycle (defined as the period from the end of one depot maintenance period to the end of the next), the *type* of maintenance needed (i.e., docking or non-docking), and the *timing of events* within the cycle affect the carrier's availability to meet operational needs.

The length of the cycle for aircraft carriers has changed several times in the last two decades. Currently, the Navy uses a 32-month cycle. This cycle has increased a carrier's ability to provide additional forward presence as requested by theater commanders (this additional presence is called "surge"). However, the combination of a 32-month cycle length with the personnel tempo policy limit of one 6-month deployment per cycle has reduced the proportion of time that a carrier is deployed. The reduction in the percentage of time that each carrier is deployed, coupled with the decrease in the number of carriers in the fleet, makes it difficult for operational planners to meet the forward-presence requirements of theater commanders.

Recognizing the challenge, the Navy asked RAND to assess the implications of different cycle lengths and their effect on the forward presence of *Nimitz*-class aircraft carriers. We assume a deployment length of six months and, in accordance with personnel policies in place under the 32-month cycle, also assume that the time between deployments will equal twice the length of the previous deployment. We assess several one-deployment cycles as well as potential two-deployment options. We also analyze the impact of different cycles on managing shipyard workloads.

Cycles and Operational Availability

Given a fixed number of months for scheduled maintenance, deployments, and the time between deployments, Navy planners face a three-sided tradeoff in setting a carrier's cycle length. They must balance the goals of

- deploying carriers and generating forward presence
- holding carriers in reserve and keeping them surge-ready to meet emerging needs
- maintaining the materiel condition of the ships.

This is a zero-sum tradeoff in which improving one goal can adversely affect the others. Under the current 32-month, one-deployment cycle, for example, a carrier is deployed 19 percent of the time, at Major Combat Operations–Surge (MCO-S)/Major Combat Operations–Ready (MCO-R) status (i.e., able to deploy within 30 days) 46 percent of the time, at Maritime Security Surge (MSS) status (i.e., able to deploy within 90 days) 11 percent of the time, and in depot maintenance 24 percent of the time. This is depicted in the third column of Figure S.1. This carrier thus contributes to the "6+1" fleet response plan goal by being available to serve as one of the six ships 65 percent of the time and as the seventh ship 11 percent of the time.[1]

---

[1] This goal is to have at least six carriers deployed (or able to deploy) within 30 days, and a seventh carrier deployed (or able to deploy) within 90 days.

Figure S.1
The Impact of Different Maintenance Cycles on the Operational Availability of a Notional Carrier

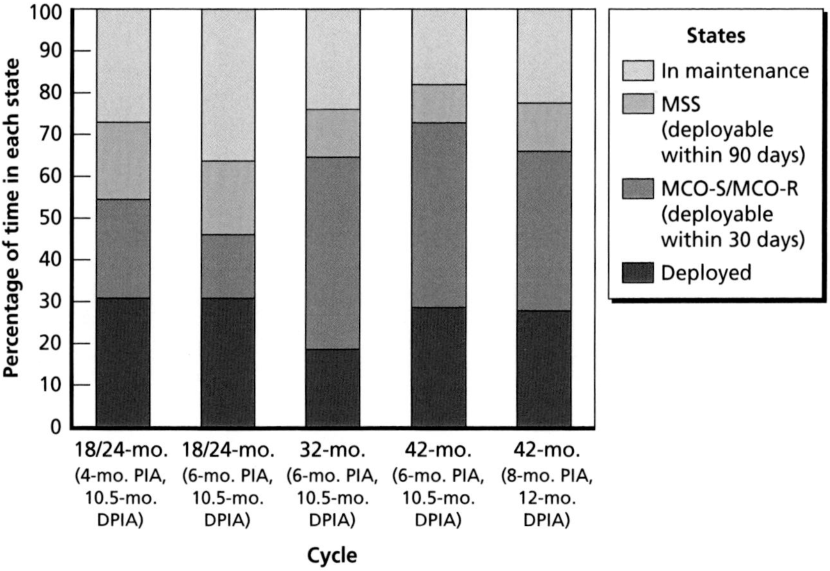

NOTE: PIA = planned incremental availability. DPIA = docking planned incremental availability.
RAND *MG706-S.1*

Reducing the cycle length to 18 months increases deployment to 31 percent of the time but decreases MSS or higher readiness to 33 percent. It also increases the time in scheduled maintenance to 36 percent of a carrier's life. This is depicted in the second column of Figure S.1.

Lengthening the cycle to 42 months and adding a second deployment in the cycle results in a carrier being deployed 29 percent of the time and at MSS or higher readiness an additional 53 percent of the time. This is depicted in the fourth column of Figure S.1. This would allow the fleet to meet the 6+1 fleet goal 100 percent of the time.

Reducing the length of PIAs for depot maintenance repair from six to four months—as may be possible under a cycle featuring 18 months between PIAs (i.e., in an 18/24-month cycle)—increases the proportion of time a ship is able to surge. This is shown in the first column of Figure S.1. Alternatively, extending the length of PIAs—as may be required under a 42-month cycle—reduces the amount of time a ship is able to surge. The fifth column of Figure S.1 shows a 42-month cycle with an 8-month depot maintenance period. Extending the maintenance period beyond a 6-month duration increases training time and decreases the amount of time a carrier is able to surge.

Cycles and Shipyard Workload

We also assessed the technical feasibility of maintenance cycles shorter or longer than the current 32-month cycle. Prior to the current 32-month cycle, *Nimitz*-class carriers operated on 24–27 month cycles. This suggests that shorter cycles, by offering more frequent opportunities to accomplish depot work, are technically feasible. Shorter cycles may also help level-load work at the shipyards, with more frequent depot visits resulting in smaller work packages.

Norfolk Naval Shipyard and Puget Sound Naval Shipyard are the two public shipyards that perform depot-level maintenance for aircraft carriers during availabilities. These shipyards can efficiently execute approximately 30,000 man-days per month during a typical availability in the 32-month, one-deployment cycle. We assume that the PIAs for the 18/24-month, one-deployment cycle would range from 15,000–25,000 man-days per month. As such, they could, perhaps, be accomplished within four months, as suggested above.

Extending the maintenance cycle beyond the current 32 months raises several questions of feasibility. Certain maintenance tasks must be performed at specified times to ensure that a carrier reaches its operational life of approximately 50 years. Some of these tasks could perhaps be performed earlier or later than currently planned; engineering studies, such as those conducted when the cycle was extended from 27 to 32 months, would be required should the Navy consider extending the cycle beyond 32 months. Some of the longer, two-deployment cycles could require that up to 375,000 man-days of work be accomplished within a 6-month availability—this amount of work is more than twice what Navy depots could be expected to accomplish in that period of time. This could require extension of the PIA beyond the nominal six months, as noted above.

Longer cycles with large work packages lead to larger peaks and deeper valleys in the carrier workload at a shipyard. These peaks and valleys make it difficult to efficiently manage the depot workforce and can lead to higher workforce costs. The longer, two-deployment cycles could result in long periods (of several months to more than a year) when there are no carriers at a shipyard for depot maintenance. These gaps could lead to a loss of learning or currency in maintenance tasks that are performed infrequently. This loss of learning could increase the size of the work packages and lead to higher costs.

Stretching depot availabilities beyond their notional lengths to handle larger workloads could help level-load the shipyard, but would also require more training (or retraining) for the ship's crew after maintenance. Extended maintenance and training would reduce the time a ship is at MCO-S or higher readiness, thereby negating a chief advantage of the longer, two-deployment cycles.

Findings and Recommendations

On balance, our analysis suggests that shortening the one-deployment cycle will increase the forward presence of the carrier fleet but reduce its ability to meet the 6+1 fleet goal. Table S.1 summarizes the advantages and disadvantages of each notional cycle examined above.

Shorter cycles may help level workloads at the shipyards. While longer, two-deployment cycles may increase forward presence while sustaining higher levels of readiness for longer periods of time, they could complicate workforce management at public shipyards. The Navy's 30-day continuous maintenance availabilities between deployments may not provide the deep maintenance needed between deployments, and a backlog of deferred work is likely to develop. Even if the carrier depot workload were to remain unchanged in the two-deployment cycle, fewer opportunities for depot maintenance would lead to larger work packages. Our workload estimates suggest that the PIA, docking planned incremental availability, and carrier incremental availability

Table S.1
The Effects of Cycles Shorter or Longer Than the Baseline 32-Month Maintenance Cycle

Metrics	Shorter Cycle (e.g., 18/24-mo., one-deployment)	Longer Cycle (e.g., 42-mo., two-deployment)
Time a carrier is deployed	Increased	Increased, if maintenance workload can be managed
Surge readiness (deployable within 30–90 days)	Decreased	Increased
Ability to meet 6+1 fleet goal	Decreased	Increased
Ability to level-load work across time at shipyards	Increased	Decreased
Maintenance demands	More frequent	May create deferred-work backlogs

work packages could grow to the point where they could not be executed in the time we assumed. The Navy could perform engineering studies to examine the impact of increased maintenance demands in two-deployment cycles.

The Navy has adjusted personnel tempo policies to better provide carriers where and when needed. Current plans to meet demands for aircraft carrier presence include extending deployment lengths, reducing turnaround times, and, in some cases, including two deployments per cycle. Deployments may be longer or shorter than six months and carriers may redeploy more quickly. Increased operational tempo may adversely affect the Navy's ability to meet the maintenance demands of the carriers and retain and recruit personnel. Our analysis offers options for increasing carrier forward presence while keeping previous personnel tempo policies intact.

Acknowledgments

We thank CAPT Charles Davis of OPNAV N81 for his guidance and support during the course of the research. CAPT Catherine Osman, CDR Jim Brown, and Nancy Harned from N81 also shared their time, insights, and suggestions during the course of the research. We gratefully acknowledge the assistance of the staff of the Carrier Planning Activity, Chesapeake, Virginia. Nick D'Amato, Valerie Howe, Brad Toncray, Lew Rankin, Gregg Baumeier, and Bob Bolden assisted us in addressing the planning needed to meet many aircraft carrier maintenance demands. Kelly Powers, NAVSEA 08 (Nuclear Propulsion), provided us with a better understanding of the maintenance demands of an aircraft carrier's nuclear propulsion system. We also appreciate the assistance provided by Dana Dervay and LCDR Dave Cimprich of Commander, United States Fleet Forces Command. We gratefully acknowledge the guidance and direction provided by VADM David Architzel, former Program Executive Officer for Aircraft Carriers, and CAPT Thomas Moore from his staff.

CAPT Ralph Soule, Ship Materiel Readiness, Commander Naval Air Forces, and Giles Smith of RAND performed technical reviews of an earlier draft of the report and provided a number of useful comments that helped strengthen the overall report. At RAND, this project benefited from the thoughts and suggestions of our colleagues Tal Manvel, Robert Murphy, Mark Arena, and John Birkler. Debbie Peetz provided overall support to the study and Vicki Wunderle provided administrative assistance.

Abbreviations

AOR	area of responsibility
CIA	carrier incremental availability
CM	continuous maintenance
CMA	continuous maintenance availability
COCOM	combatant command
COH	complex overhaul
COMPTUEX	Composite Training Unit Exercise
CPA	Carrier Planning Activity
CSG	carrier strike group
CVN	aircraft carrier, nuclear
DPIA	docking planned incremental availability
DSRA	docking selected restricted availability
EOC	engineered operating cycle
FDNF	Forward Deployed Naval Forces
FRP	Fleet Response Plan
FY	fiscal year
IMP	Incremental Maintenance Program

JTFEX	Joint Task Force Exercise
MCO-R	Major Combat Operations–Ready
MCO-S	Major Combat Operations–Surge
MSS	Maritime Security Surge
NAVSEA	Naval Sea Systems Command
NAVSEA 08	Nuclear Propulsion Directorate of the Naval Sea Systems Command
NNSY	Norfolk Naval Shipyard
OPNAV N81	Assessments Division of the Deputy Chief of Naval Operations for Resources, Requirements, and Assessments
PERSTEMPO	personnel tempo
PIA	planned incremental availability
PSA	post-shakedown availability
PSNSY	Puget Sound Naval Shipyard
RCOH	refueling complex overhaul
SRA	selected restricted availability
SRF	ship repair facility
TAR	turnaround ratio
WARR	workload allocation and resource report

CHAPTER ONE

Introduction

Background

The U.S. Navy currently maintains a fleet of 11 aircraft carriers. These ships, which are among the most powerful and versatile elements of U.S. naval forces, allow the Navy to undertake a wide variety of tasks. These tasks include bringing airpower to bear against opponents, deterring adversaries, engaging friends and allies, providing humanitarian assistance, and other, evolving missions the military is likely undertake in coming years.[1]

Carriers, like all U.S. Navy ships, operate on a cycle that includes training to achieve readiness goals and then sustaining high readiness levels for a period of time. A deployment to a forward theater of operations is part of the readiness sustainment cycle. At the end of the training–readiness–deployment period, the ship enters a shipyard for depot-level repair and modernization work; this period is called an "availability."

Carriers are large, complex systems whose crew require extensive training and practice in the operations and safety of the ship, the integration of the ship and the air wing, and the integration of all the ships in the carrier strike group (CSG). Because of their complexity, the ships themselves require a great deal of maintenance. Hence, there

[1] John Gordon IV, Peter A. Wilson, John Birkler, Steven Boraz, and Gordon T. Lee, *Leveraging America's Aircraft Carrier Capabilities: Exploring New Combat and Noncombat Roles and Missions for the U.S. Carrier Fleet*, Santa Monica, Calif.: RAND Corporation, MG-448-NAVY, 2006.

is a tradeoff between the cycle length and the proportion of time a carrier is deployed or available to deploy. With just one deployment per cycle, longer cycles reduce the proportion of time a carrier is deployed, but can increase the proportion of time the carrier is not in scheduled maintenance and is able to respond to contingencies and crises.

The cycle for aircraft carriers has changed several times in the last two decades. The introduction of the Incremental Maintenance Program (IMP) for *Nimitz*-class carriers in 1994 set the cycle length at 24 months. The Fleet Response Plan (FRP) extended the cycle length to 27 months in 2003. In August 2006, the cycle length was extended to 32 months.[2]

The Challenge

Increasing the length of the carrier cycle from 27 to 32 months has increased the "surge" readiness of the carrier fleet, but, given the limit of one 6-month deployment per cycle, has reduced the proportion of time the carrier is deployed. This lengthened 32-month carrier cycle, coupled with the recent reduction in the size of the carrier fleet from 12 to 11 ships, had made it difficult for the Navy to meet the forward presence requirements of theater commanders. This challenge will increase in the four years between 2013—when the USS *Enterprise* is decommissioned (reducing the fleet size to ten carriers)—and 2017, when the USS *Gerald R. Ford*, the first of a new class of carriers, becomes operationally available.

Recognizing this problem, the Assessments Division of the Deputy Chief of Naval Operations for Resources, Requirements, and Assessments (OPNAV N81) asked RAND to examine the feasibility and implications of cycles that would increase the percentage of time that a carrier is deployed. Of particular interest are cycles that would permit a

[2] Department of the Navy, OPNAV Notice 4700, "Representative Intervals, Durations, Maintenance Cycles, and Repair Mandays for Depot Level Maintenance Availabilities of U.S. Navy Ships," August 31, 2006b.

carrier to perform two deployments between major depot availabilities. Our study addresses

- how the duration and schedule of events within a cycle could be changed to provide greater forward presence while maintaining high readiness rates
- whether such cycles are technically feasible from the perspective of accomplishing required maintenance
- the impact of varying the cycle length on operational availability
- the impact of varying the cycle length on the maintenance industrial base, including the cost of conducting depot-level maintenance.

Analytical Approach

To address these issues, we first defined new cycles that could increase the percentage of time that a carrier is deployed, focusing on the maintenance of *Nimitz*-class carriers. Then, working closely with the Carrier Planning Activity (CPA) and the Nuclear Propulsion Directorate of the Naval Sea Systems Command (NAVSEA 08), we assessed the ability of each cycle to meet maintenance requirements. Using several analytical tools developed during the course of the research, we estimated the effects of each cycle on various measures of operational availability, the workload demands placed on the maintenance industrial base, and the cost of providing depot-level maintenance.

Several issues relevant to the setting of carrier deployments and cycle lengths were beyond the scope of the research. Specifically, we did not examine

- **The impact of increased deployments on the operational life of the nuclear fuel in the carrier's reactors.** Currently, *Nimitz*-class carriers are scheduled for a midlife refueling complex overhaul (RCOH) after approximately 23 years. Increased deployments could deplete reactor fuel sooner than expected, require refueling sooner than planned, and shorten the planned 50-year life of the *Nimitz*-class ships. These contingencies could affect the size of the fleet and its ability to pursue national interests.

- **The availability of air wings to meet carrier deployment schedules.** Naval aircraft follow their own cycles of training, readiness, and maintenance events. We assumed that ready air wings would be available to support carrier deployments.
- **Possible changes in CSG training demands and schedules.** Different cycles, especially cycles that involve two deployments, may require different training strategies. We did not examine training events and schedules that would potentially change as cycle lengths or number of deployments within a cycle change.

Organization of the Monograph

In Chapter Two, we provide an overview of the aircraft carrier fleet and the past and current maintenance cycles of *Nimitz*-class carriers. We also define three new cycles—one shorter, one-deployment cycle and two longer, two-deployment cycles—and discuss their technical feasibility. In Chapter Three, we describe the impact of varying cycles on measures of operational availability. In Chapter Four, we consider the impact of varying cycles on depot workforce management and on the cost of providing depot-level maintenance to the carrier fleet. In Chapter Five, we offer our conclusions and recommendations.

CHAPTER TWO

Past, Current, and Potential Carrier Cycles

Over the next two decades, the number of aircraft carriers in the Navy's fleet will vary between 10 and 12. The ability of these carriers to deploy or be deployed will, as noted, depend in part on their operational and maintenance cycles. Below, we describe the current and planned fleet of carriers. We then discuss the evolution of maintenance cycles for *Nimitz*-class carriers, including how their maintenance policies and operational cycles have varied. We also discuss some potential cycles for evaluation. The technical feasibility of these cycles is considered in subsequent chapters.

The U.S. Carrier Fleet

Table 2.1 lists current and planned vessels in the U.S. carrier fleet. The USS *Kitty Hawk*, the only operational non-nuclear carrier, is based in Japan as part of the Forward Deployed Naval Forces (FDNF).[1] She will

[1] FDNF aircraft carriers are maintained on a different schedule from carriers based in the United States. Forward-presence requirements dictate shorter but more frequent maintenance availabilities. The normal schedule calls for annual 4-month maintenance availability, from January to May, performed in Japan using local shipyards for non-nuclear work, and workers from Puget Sound Naval Shipyard for nuclear work. FDNF carrier maintenance planning was not included in this study, but recent engineering studies regarding a nuclear-powered FDNF carrier may provide insights about the benefit of a shortened operational cycle for *Nimitz*-class carriers. We later discuss the operational need for six carriers to be deployed or able to deploy within 30 days at any given time; given its forward presence and its peculiar maintenance needs and schedule, the FDNF carrier is always counted as one of these six carriers.

Table 2.1
Current and Planned U.S. Navy Aircraft Carrier Fleet

Aircraft Carrier	Hull Number	Year Commissioned	Expected Retirement	Homeport
USS *Kitty Hawk*	CV 63	1961	2008	Yokosuka, Japan
USS *Enterprise*	CVN 65	1961	2013	Norfolk, Va.
USS *Nimitz*	CVN 68	1975	2027	San Diego, Calif.
USS *Dwight D. Eisenhower*	CVN 69	1977	2029	Norfolk, Va.
USS *Carl Vinson*	CVN 70	1982	2034	Norfolk, Va.[a]
USS *Theodore Roosevelt*	CVN 71	1986	2038	Norfolk, Va.
USS *Abraham Lincoln*	CVN 72	1989	2041	Everett, Wash.
USS *George Washington*	CVN 73	1992	2044	Norfolk, Va.
USS *John C. Stennis*	CVN 74	1995	2047	Bremerton, Wash.
USS *Harry S. Truman*	CVN 75	1998	2050	Norfolk, Va.
USS *Ronald Reagan*	CVN 76	2003	2055	San Diego, Calif.
USS *George H. W. Bush*	CVN 77	2008	2060	East Coast
USS *Gerald R. Ford*	CVN 78	2015	2067	West Coast
CVNX2	CVN 79	2019	2071	East Coast
CVNX3	CVN 80	2025	2077	West Coast

NOTE: CVN = aircraft carrier, nuclear.

[a] As of fall 2007, the *Vinson* is at Northrop Grumman Newport News for her midlife RCOH. She will be homeported at San Diego following her RCOH.

be decommissioned in 2008, when the USS *George Washington* will replace her as the FDNF carrier. The USS *George H. W. Bush*, the tenth and last of the *Nimitz*-class carriers, will be commissioned in 2008.

Current plans call for decommissioning the USS *Enterprise,* currently the only non-*Nimitz* class nuclear aircraft carrier, in 2013. Between her decommissioning and the commissioning in 2015 of the USS *Gerald R. Ford,* the carrier fleet will diminish to ten ships. Because the *Ford* will require some 30 months to become operationally ready to deploy after commissioning, this operational gap may be even longer. This gap will severely strain the Navy's ability to meet the forward-presence requirements of theater commanders.

Initial Maintenance Cycles for *Nimitz*-Class Carriers

When the USS *Nimitz* entered service in 1975, nuclear carriers followed the engineered operating cycle (EOC) developed for conventional carriers. This cycle included an 18-month period for training and deployment followed by a depot availability. The length of time and number of man-days needed to accomplish the workload of the depot periods grew as the carrier aged. A 3-month selected restricted availability (SRA) followed the first operational period, and a 5-½-month docking selected restricted availability (DSRA) followed the second operational period. A second, 3-month SRA followed the third operational period, and an 18-month complex overhaul (COH) followed the fourth operational period. This SRA–DSRA–SRA–COH cycle was then repeated, with the second COH lasting 24 months (see Figure 2.1).

The EOC resulted in operational and funding problems for the *Nimitz* class. The concentration of work in the COHs resulted in a period of nearly two years during which a carrier was unavailable for training or operations. The long maintenance period resulted in high crew turnover, requiring significant training and retraining of the crew and making it difficult to achieve adequate levels of crew proficiency in time for scheduled deployments. The large amount of work in the COH strained the ability of the maintenance industrial base to com-

Figure 2.1
Comparison of EOC and IMP Cycles for a Notional Nuclear Carrier

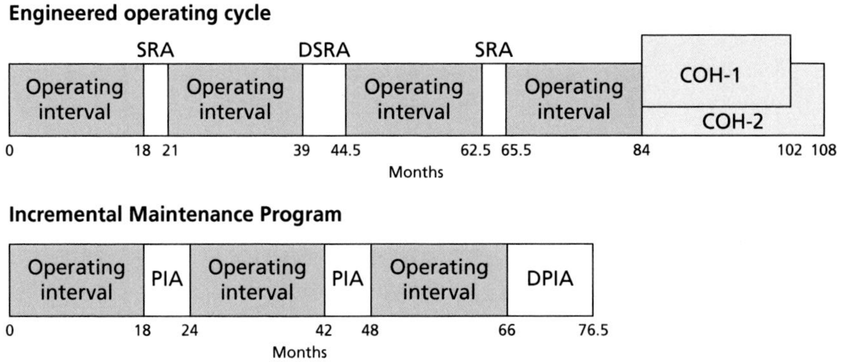

NOTE: An operating interval generally includes a deployment.
RAND *MG706-2.1*

plete the required maintenance and modernization tasks in the time allocated. It also required high levels of funding to pay for the work.

To address these problems, the Navy instituted the IMP in 1994. The IMP retained the 18-month operational periods while spreading the depot maintenance workload more evenly over the life of the ship. Six-month planned incremental availabilities (PIAs) followed the first and second operational periods, and a 10-½-month docking planned incremental availability (DPIA) followed the third operational period. The PIA–PIA–DPIA sequence was then repeated. An RCOH that lasted approximately three years followed the third PIA–PIA sequence (at approximately the 23-year point). The sequence was repeated over the second half of the ship's life. Figure 2.1 compares the EOC and IMP cycles.

Like the EOC, the IMP recognized that maintenance requirements increased as the carrier aged. The notional depot-level man-days for the second set of depot availabilities (called PIA2s and DPIA2s) were about 15 percent higher than the notional man-days for the first set (called PIA1s and DPIA1s) and about 15 percent less than the man-days for the third set (called PIA3s and DPIA3s). Still, the duration of the PIAs was held constant at six months and that of the DPIAs at 10-½ months. Also, the PIA–PIA–DPIA sequence started with PIA2s and DPIA2s after the midlife RCOH (i.e., the notional PIA1 and DPIA1 work packages only applied to the first six years of a carrier's operational life).

Under the original IMP with its 24-month PIA maintenance cycles, there were 12 operating intervals, and therefore 12 deployments, for *Nimitz*-class carriers both before and after the midlife RCOH. The IMP dampened the funding spikes required under the EOC and spread depot-level maintenance work more evenly over time. The IMP also helped maintain better overall ship conditions, resulting in a higher degree of material readiness for the carrier fleet. The percentage of time that a carrier was in maintenance and the percentage of time it was deployed changed little from the EOC to the IMP.

Introduction of the Fleet Response Plan

One shortcoming of EOC and IMP's 24-month cycle was the cycle's inefficient use of crew readiness levels. Carriers returning from deployment are at their highest state of readiness, having conducted operations for several months. Yet under the EOC and IMP cycles, a carrier stood down almost immediately after deployment, entering its depot maintenance availability and sending its crew on post-deployment leave. Little training was accomplished during the stand-down and maintenance periods. As a result, the carrier's training readiness level was low before, during, and upon leaving maintenance. Readiness levels increased as the ship accomplished its required training tasks and until the carrier was ready for its 6-month deployment (approximately 12 months after the end of the last depot period and six months before its next one).

To increase overall readiness of the carrier fleet to respond to contingencies and crises, the Navy implemented the FRP in 2003. Under the FRP, a carrier attains readiness sooner and sustains it longer.

The FRP cycle normally begins with basic training. The goal of basic training is to ensure that the crew can safely operate the ship, is ready to support equipment testing, and is qualified for underway watch stations. Basic training occurs both during the maintenance period, with team training ashore and onboard, and after the ship leaves the depot.[2] Upon completion of basic training, a carrier's crew achieves prescribed certifications, is proficient in Navy Mission Essential Tasks, and is ready for integrated training events.[3] Carriers that complete basic training may be tasked with operations consistent with their level of training. These carriers attain Maritime Security Surge (MSS) status, meaning that they can be made ready for a surge deployment within 90 days.[4]

Once basic training is completed, integrated phase training begins. The goal of the integrated phase is to bring together the individual ships in the CSG to allow group-level training and operations in a challenging environment. Integrated training can be tailored to meet the specific needs of a combatant command (COCOM). A carrier remains in integrated training for approximately three months.

Two major underway-training events occur after basic phase training: a Composite Training Unit Exercise (COMPTUEX) and a Joint Task Force Exercise (JTFEX). Upon completing a COMPTUEX, a ship becomes Major Combat Operations–Surge (MCO-S) ready, meaning that it can be made able to deploy within 30 days. Upon completing a JTFEX, which usually occurs about three months after it completes basic training, a ship is Major Combat Operations–Ready (MCO-R). MCO-R is the readiness goal for all deploying CSGs, and means that the CSG is ready and certified for all forward-deployed operations.[5]

The sustainment phase begins after the completion of integrated training. The ship sustains its high readiness level for a period of approximately 12 months, including a 6-month deployment. At the comple-

---

[2] The length of basic training depends on the time spent in depot maintenance. Longer maintenance periods require longer training periods due to increased crew turnover during maintenance.

[3] Department of the Navy, Chief of Naval Operations Instruction 3000.15, "Fleet Response Plan," August 31, 2006a.

[4] Once a carrier completes maintenance and starts basic training, it is considered an asset that can be deployed in a crisis situation. If the need arises, basic training can be accelerated to meet surge demands. MSS status was formerly called *Emergency Surge Ready status*.

[5] MCO-S was formerly called *Surge Ready*; a ship that is MCO-R was formerly considered to have reached *Routine Deployable status*.

tion of deployment, a carrier remains in the sustainment phase and is a deployable asset until the start of its next maintenance period.

Recent Changes to the FRP Cycle Length

In addition to placing increased emphasis on training and the sustainment of readiness, the FRP lengthened the carrier cycle from the notional 24 months of the IMP to 27 months. The FRP did not change the 6-month length of PIAs or the 10-½-month length of DPIAs. The 27-month cycle formalized what had already evolved in practice. Figure 2.2 shows the average elapsed time between the start of depot availabilities for *Nimitz*-class carriers. For the eight carriers shown in Figure 2.2, the average time between depot availabilities regularly exceeded 24 months. The average time for three carriers exceeded 27 months.

Recently, the FRP cycle length was increased to 32 months. The new cycle results in eight years (96 months) between docking avail-

Figure 2.2
Average Number of Months Between Start of *Nimitz*-Class Depot Availabilities (1977–2005)

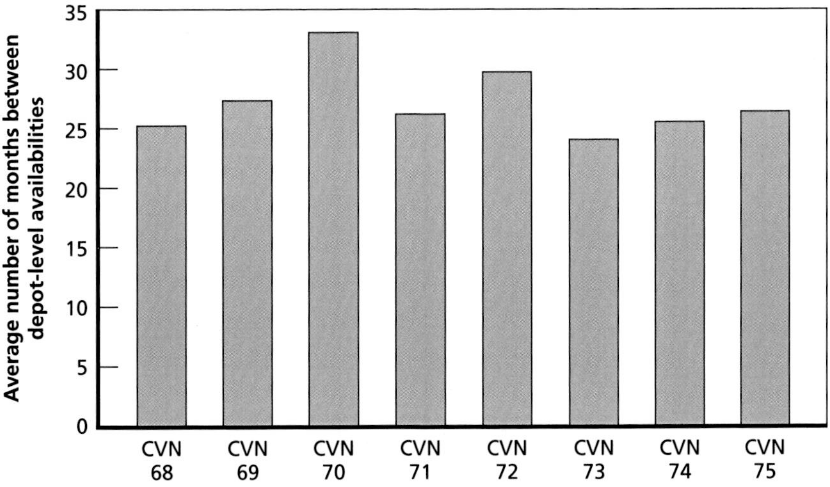

RAND *MG706-2.2*

abilities.[6] The increase in cycle length, coupled with FRP policies, has had several effects. The percentage of time that a carrier is deployable grows as cycle length increases, because the percent of time the carrier spends in maintenance decreases. Yet because there has been only one deployment per cycle, the percentage of time that a carrier is actually deployed decreases as the cycle length grows. This reduction, coupled with a shrinking carrier force, presents challenges to the Navy in meeting forward-presence requirements.

Meeting Forward-Presence Demands

The operational scheduling and deployment of carriers are dictated by the need of COCOMs for carrier presence in their areas of responsibility (AORs). Demands for carrier presence change over time and are based on the demands of U.S. national interests. The FRP was designed to enable the carrier fleet to meet these changing demands.

Fleet schedulers must balance the maintenance, training, deployment, and readiness sustainment of carriers to meet presence demands. They must also consider the overall goal of a "6+1 fleet" that has at least six carriers deployed (or able to deploy) within 30 days, and a seventh carrier deployed (or able to deploy) within 90 days. Schedulers begin the scheduling process by laying out the notional carrier maintenance periods. They then schedule the appropriate time for basic phase training (normally three months) and integrated training (an additional three months) for each carrier. These periods in turn dictate the sustainment period of a ship before, during, and after deployment.

The greatest challenge to scheduling is meeting a COCOM's demand for an additional carrier in an AOR. Schedulers must then evaluate each carrier's level of training and readiness and decide which carrier can respond. This may be a carrier completing a deployment and

---

[6] In an effort to reduce maintenance costs, the Chief of Naval Operations approved a 12-year docking interval for the USS *Nimitz*. If that interval proves feasible from a required maintenance perspective, then the docking intervals for all *Nimitz*-class carriers after their midlife RCOH will be extended to 12 years. The CPA estimates that the resulting DPIA will be 15 months long and require 500,000 man-days of work.

returning to its homeport, or a carrier that is next scheduled to deploy. These scheduling challenges are increasing as the number of operational carriers decreases but tensions increase throughout the world.

Potential Cycles for Evaluation

Our research goal was to formulate new carrier cycles that would maintain or increase forward presence while maintaining high levels of surge capability. We focused on cycles that would allow a carrier to make two deployments between major depot availabilities. We also examined one-deployment cycles that were shorter than the current 32 months.

In developing the new cycles, we assumed the following:

- The duration of PIAs and DPIAs remains fixed at six and 10-½ months, respectively. These durations encompass the period of time between a carrier's entrance and departure from a depot.
- The time between dockings (i.e., DPIAs) does not exceed 12 years.
- The time required for basic training is three months following a PIA, five months following a DPIA, seven months following an RCOH, and nine months following the post-shakedown availability (PSA) that accompanies the introduction of each new carrier.[7] The duration and intensity of basic training depend on the degree of crew turnover during the maintenance period and the need for the crew to perform maintenance tasks that hinder their availability for training; longer availabilities require longer basic training periods. If needed, ships can receive tailored training to meet a surge capability and be deployed within 90 days.
- Ships attain MCO-R status at the completion of integrated training, which lasts three months.

---

7 When a new carrier is delivered to the Navy, the crew conducts a shakedown cruise to identify any manufacturing or equipment problems. The shipbuilder, Northrop Grumman Newport News, corrects these problems during the PSA. At the conclusion of the PSA, the carrier crew commences training for its first operational deployment.

- Deployments last six months.
- A minimum of 12 months between deployments is required to maintain a turnaround ratio (TAR) of 2.0 because current personnel tempo (PERSTEMPO) policies stipulate that a ship cannot deploy again until twice the time spent on the previous deployment has elapsed.

Figure 2.3 shows the cycles we considered in this study. The development of three of these cycles—the 24-, 27-, and 32-month cycles—is described above. The following paragraphs present summaries of these and three other potential cycles—an 18/24-month, one-deployment cycle; a 36/42-month, two-deployment cycle; and a 42-month, two-deployment cycle. Because the length of a maintenance cycle is measured from the beginning of one maintenance period to the beginning of the next, all the maintenance cycles we consider begin with a maintenance period. Figure 2.3 also shows periods in each cycle for

- basic training (MSS), when a carrier can be made ready to deploy within 90 days
- extended training (MCO-S), when a carrier can be made ready to deploy within 30 days
- Routine Deployable status (MCO-R), when a carrier is ready for but not on deployment
- deployment.

An *18/24-month* cycle is the shortest possible maintenance cycle for aircraft carriers, given our assumptions. It would feature six months for PIA maintenance, six months total for basic, integrated, and sustainment training, and six months for deployment. The cycle is extended to 24 months when beginning with DPIA maintenance, including 10-½ months for maintenance and 7-½ months for training. Note that ships on this cycle immediately deploy after completing training and start depot maintenance immediately after deployment. Furthermore, ships on this cycle would feature 96 months between dockings, the same as the 32-month cycle most recently used. This cycle option maximizes deployment over deployability.

Figure 2.3
Alternative One-Deployment Cycles

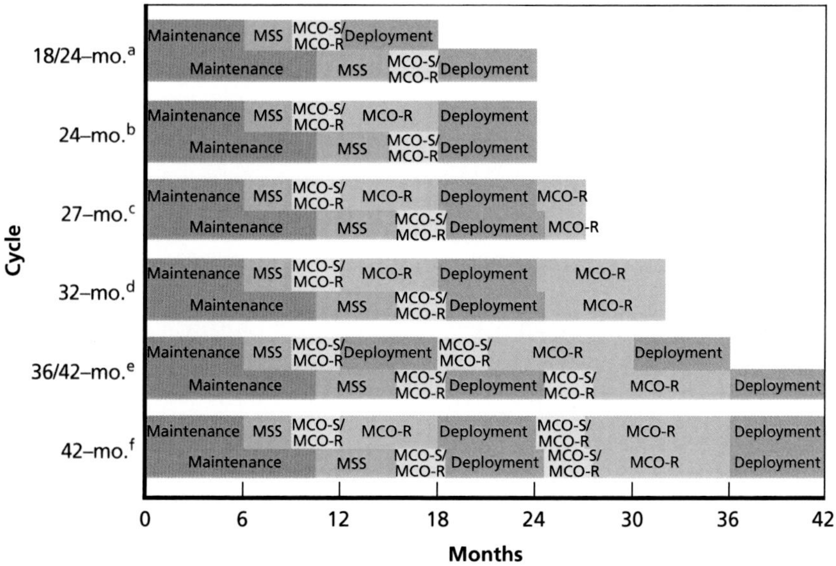

[a]PIA–PIA–PIA–PIA–DPIA; 96 months between dockings.
[b]PIA–PIA–DPIA; 72 months between dockings.
[c]PIA–PIA–DPIA; 81 months between dockings.
[d]PIA–PIA–DPIA; 96 months between dockings. This is the current cycle.
[e]PIA–PIA–DPIA; 114 months between dockings.
[f]PIA–PIA–DPIA; 126 months between dockings.

RAND *MG706-2.3*

The *24-month* cycle is the same cycle the Navy used when first introducing the IMP. Cycles beginning with a PIA include a period when a ship is deployable but not actually deployed. Cycles beginning with a DPIA effectively trade this deployability time for more maintenance.

The *27-month* cycle is the cycle the Navy used when first implementing the FRP. Cycles beginning with a PIA include two periods in which a ship is deployable but not actually deployed; these periods occur before and after its single 6-month deployment. Cycles beginning with a DPIA effectively trade the first MCO-S period for additional maintenance.

The *32-month* cycle is the cycle currently used by the Navy. Cycles that begin with a PIA include two periods in which a ship is deployable but not actually deployed. Cycles beginning with a DPIA effectively trade the first deployability period and part of the second for more maintenance.

The *36/42-month* cycle is the shortest two-deployment cycle possible, given our assumptions about TAR and deployment length. The 36-month cycle begins with a 6-month PIA, followed by six months of training, six months for the first deployment, 12 months between deployments, and a second 6-month deployment. The 42-month cycle begins with a DPIA, extending maintenance and crew training time by six months while maintaining the length of other phases of the cycle.

The *42-month cycle* is an alternative two-deployment cycle that holds constant the time between the start of depot maintenance availabilities. It combines some of the additional deployment available in a 36-month, two-deployment cycle with the additional deployability available in a longer cycle. It differs from the 36/42-month cycle principally in that it provides an additional period of ready deployability before its first PIA cycle.

Figure 2.4 presents another perspective on these cycles. It shows the sequence of PIAs and DPIAs throughout the first half of the life of a *Nimitz*-class carrier, beginning with the end of the ship's PSA. Following the policy for the current 32-month cycle, the first DPIA for the 36/42- and the 42-month cycles are DPIA2s. The current 32-month cycle has one carrier incremental availability (CIA) period between depot availabilities.[8] We assume that the 36/42- and 42-month cycles would have two CIA periods between deployments. The tan portion of each bar shows when a ship is out of maintenance (i.e., in training, ready to deploy, or deployed). As previously noted, longer cycles feature a lower proportion of time in maintenance. The 42-month, two-deployment cycle would see a carrier in maintenance only 18 percent

---

[8] The month-long availabilities conducted at the operating base were formerly known as continuous maintenance availabilities (CMAs). Department of the Navy, OPNAV Notice 4700, "Representative Intervals, Durations, Maintenance Cycles, and Repair Mandays for Depot Level Maintenance Availabilities of U.S. Navy Ships," August 31, 2007c, changed the names of the CMAs to *CIAs*. We use this new term throughout our report.

Figure 2.4
Notional *Nimitz*-Class Maintenance Cycles Before RCOH

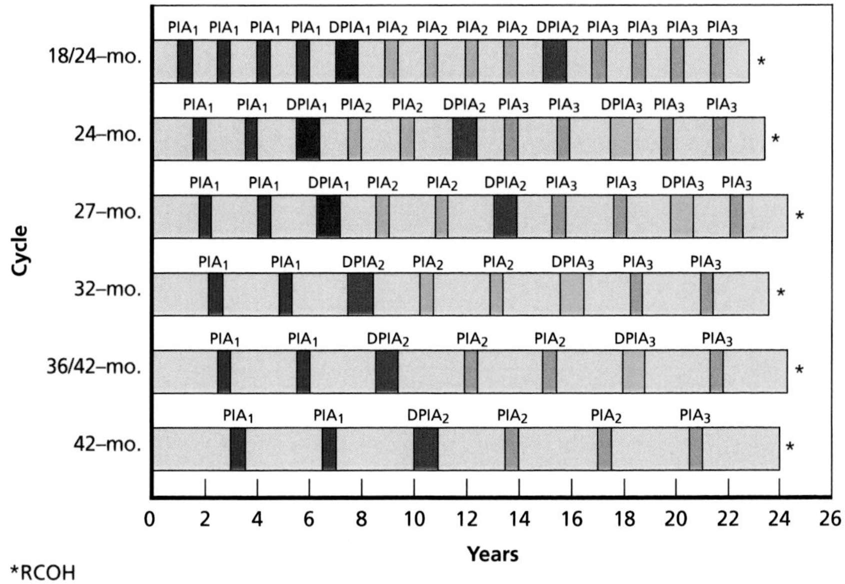

*RCOH

RAND *MG706-2.4*

of the time, while the 18/24-month, one-deployment cycle would see it in maintenance 36 percent of the time. We explore the reasons for this, as well as for the greater time spent in deployment for the two-deployment cycles and the shorter one-deployment cycles, in the next chapter.

Recent Navy Decisions to Meet Presence Requirements

We assume that the Navy will face the following constraints on its ability to meet evolving presence demands:

- force structure (number of carriers)
- duration of maintenance (normally six or 10-½ months)
- training time needed to meet deployment readiness goals (normally six months)

- deployment duration (historically six months)
- personnel policies pertaining to quality-of-life measures such as deployment versus homeport time (historically a TAR of 2.0).

Our options focus on holding these variables constant while changing the duration of the maintenance cycle.

As we conducted this research, the Navy sought to meet increased requirements for forward presence by changing the Personnel Tempo Operations Program and the duration of some deployments.[9] PERSTEMPO is defined as a unit's time away from homeport. Recent PERSTEMPO policy relaxes previous guidance and allows for increased time away from homeport, specifically through an increase in the allowable deployment duration as well as a decreased TAR (the decreased TAR still assures sailors as much time at home as deployed). While these changes will allow greater flexibility in meeting forward-presence demands, they may adversely affect quality of life for sailors. The long-term impact of this guidance on personnel retention is still undetermined. Table 2.2 provides a summary of the old and new PERSTEMPO guidance.[10]

These changes were made while holding the cycle length constant at 32 months. Our analysis, which fixes deployment length at six months and TAR at 2.0, and varies cycle length and the number of deployments, could be considered as an alternative to the new PERSTEMPO policy, or as an option for increasing carrier forward presence with traditional deployment policies.

---

9 Jack Dorsey, "Navy Changes Deployment Terms for First Time in 22 Years," *Norfolk Virginian-Pilot*, March 9, 2007.

10 For the old PERSTEMPO Program, see Department of the Navy, OPNAV Notice 3000.13B, "Personnel Tempo of Operations," February 11, 2000. For the new PERSTEMPO Program, see Department of the Navy, OPNAV Notice 3000.13C, "Personnel Tempo of Operations Program," January 16, 2007a.

Table 2.2
Summary of Changes to the U.S. Navy's PERSTEMPO Program

	Old PERSTEMPO Guidance	New PERSTEMPO Guidance
Deployment length	Maximum of six months, portal to portal; units away from homeport for greater than 56 days are considered deployed	Maximum of seven months for a single deployment within an FRP cycle; six months for units with multiple deployments within an FRP cycle
TAR	Minimum ratio of 2.0:1	Minimum ratio of 1.0:1
Homeport time	Minimum of 50 percent time in homeport over a five-year cycle (three years back and two years projected forward)	Minimum of 50 percent time in homeport over an FRP cycle

Technical Feasibility of the Potential Cycles

A key issue regarding cycles of differing length is the feasibility of maintenance within those cycles: Can the timing of the specific maintenance tasks scheduled for the various PIAs and DPIAs under the 32-month cycle be adjusted to fit within a new cycle, especially one that is longer? In an 18/24-month, one-deployment cycle, PIAs will occur sooner and more frequently, and DPIAs will occur at the same intervals as they do in the current 32-month cycle. The primary concern is whether tasks scheduled for PIAs or DPIAs that occur every 32 months can be extended for the longer, two-deployment cycles.

OPNAV Notice 4700, "Representative Intervals, Durations, Maintenance Cycles, and Repair Mandays for Depot Level Maintenance Availabilities of U.S. Navy Ships," allows a 3-month deviation from planned schedules to accommodate a ship's employment schedule or to accommodate shifts in workloads at maintenance depots. In fact, this deviation was frequently used to extend the original 24-month cycle to 27 months (see Figure 2.2). It may not be difficult to extend the tasks for the current 32-month cycle to the 36-month interval between PIAs in the 36/42-month cycle. CIAs could help accomplish tasks with hard, or mandatory, time limits of less than 36 months. The two-deployment cycles we consider have two CIAs within each cycle, providing some opportunities for accomplishing tasks with hard time limits.

Of bigger concern are (1) the extension of the docking interval in the 36/42-month cycle from 8 to 9-½ years and (2) the extension of the time between PIAs and DPIAs in the 42-month cycle. The 42-month cycle includes only one docking in each half of the carrier's life, with dockings occurring every 10-½ years. Tasks scheduled for the DPIAs in the 36/42- and 42-month cycles, which require the carrier to be in a dry dock, must be accomplished; they cannot be deferred to the next docking.

With assistance from CPA and NAVSEA 08, we explored the feasibility of extending maintenance tasks in conjunction with cycle lengths. Engineering studies—such as those conducted in preparation for the change from a 27-month to a 32-month cycle—will be required to concretely determine the viability of this option. Although the prevailing opinion is that the maintenance community can find a way to make two-deployment strategies feasible, we could not reach a definitive conclusion about the practicality of extending the timing of all required maintenance tasks to fit the longer, two-deployment cycles. This is an area ripe for further study by engineering organizations, such as CPA, that are well-attuned to the maintenance requirements of aircraft carriers. While CIA periods may help, they may require more than the one month we assume for them. PIAs and DPIAs may also need to be extended.[11] We address this issue later in the monograph.

[11] CPA already projects a 15-month DPIA when stretching the time between a carrier's RCOH and final docking to 12 years.

CHAPTER THREE

The Impact of Different Cycles on Operational Availability

The six cycles we examine have various effects on the operational availability of the fleet. In particular, as cycle length increases, so does deployability; however, the proportion of time deployed decreases. In this chapter, we examine the effects of each cycle on operational availability, first addressing the metrics for a notional carrier and then for the current and planned carrier fleet.

Relationship Between Cycle Length and Operational Readiness

Given the assumptions presented in Chapter Two, there is a direct relationship between cycle length and the percentage of time that a carrier is either deployed or capable of deploying in 30 days (this status represents the "6" in the FRP's goal of a 6+1 fleet). Figure 3.1 shows this relationship for a notional carrier operating on one-deployment cycles, assuming that the cycle length is the same for a 6-month PIA and a 10-½-month DPIA. Note that as the cycle length increases from 24 to 32 months, the proportion of time that a carrier is forward-deployed decreases from 25 to 19 percent of its operational life, while the proportion of time that it spends at MCO-S or MCO-R status increases from 30 to 46 percent. Overall, increasing the maintenance cycle from 24 to 32 months has increased the number of carriers deployed or able to deploy in 30 days by approximately 15 percent.

Figure 3.1
Percentage of Time a Notional Carrier Is Deployed or Deployable in 30 Days by Cycle Length (One-Deployment Cycles Only)

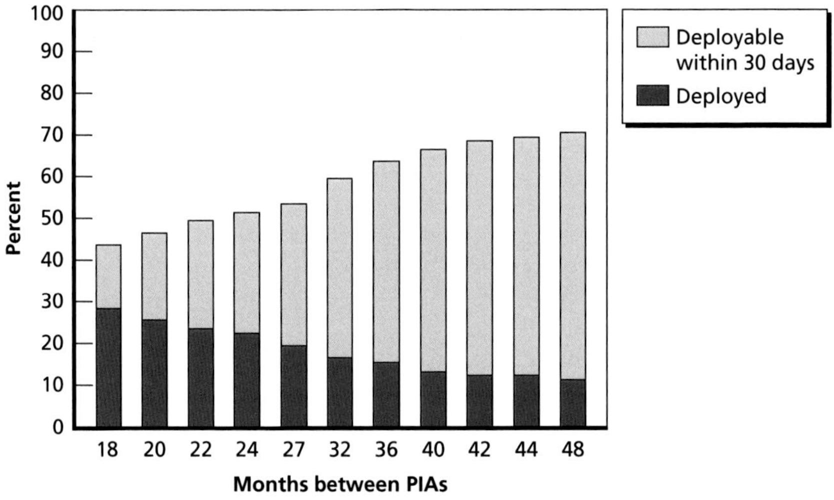

RAND *MG706-3.1*

Figure 3.2 shows the relationship between cycle length and the percentage of time that a notional carrier is deployed or capable of deploying within 30 days for two-deployment cycles. Again, the proportion of time a carrier is deployed decreases as the cycle is lengthened, while the proportion of time it would be capable of deploying within 30 days increases. The increase in deployability results from the decreased proportion of time a carrier spends in depot maintenance as the cycle length increases. As a two-deployment cycle increases from 36 to 48 months, the proportion of time a carrier is deployed, fixed at two 6-month deployments per cycle, decreases from 33 to 25 percent. Yet the proportion of time it is otherwise deployable in 30 days—equal to the difference between the total cycle length and the combined fixed amounts of time it is deployed, in depot maintenance, or in initial training after depot maintenance—increases from 36 to 48 percent.

Figure 3.3 combines the various readiness metrics for one- and two-deployment cycles. It shows, by cycle length, the proportion of time a carrier is deployed (represented by solid lines) or is able to deploy in no more than 90 days (i.e., at MSS status or higher, represented

Figure 3.2
Percentage of Time a Notional Carrier Is Deployed or Deployable in 30 Days by Cycle Length (Two-Deployment Cycles Only)

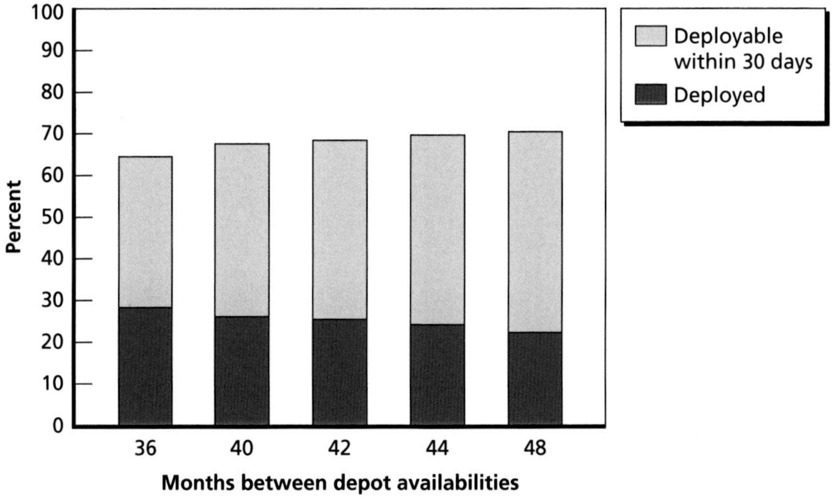

RAND *MG706-3.2*

by dashed lines). It indicates that a carrier can achieve the same level of forward presence on an 18-month, one-deployment cycle as on a 36-month, two-deployment cycle.

Carriers with one-deployment cycles spend less time at MSS or higher readiness status because they are in depot maintenance availabilities more often. For example, a ship on an 18/24-month, one-deployment cycle will spend 36 percent of its time in maintenance (see Table 3.1, which summarizes the various measures of operational readiness for the six cycles we evaluate). By contrast, a ship on a 36/42-month, two-deployment cycle will spend only 20 percent of its time in maintenance. This leads to large differences in the time a ship is not deployed and not in maintenance, and to especially large differences in the proportion of time a ship is able to deploy in no more than 30 days (i.e., time it spends in MCO-S or MCO-R status).

Figure 3.3
Percentage of Time a Notional Carrier Is Deployed or Deployable in 90 Days or Better (MSS+) by Cycle Length and Number of Deployments per Cycle

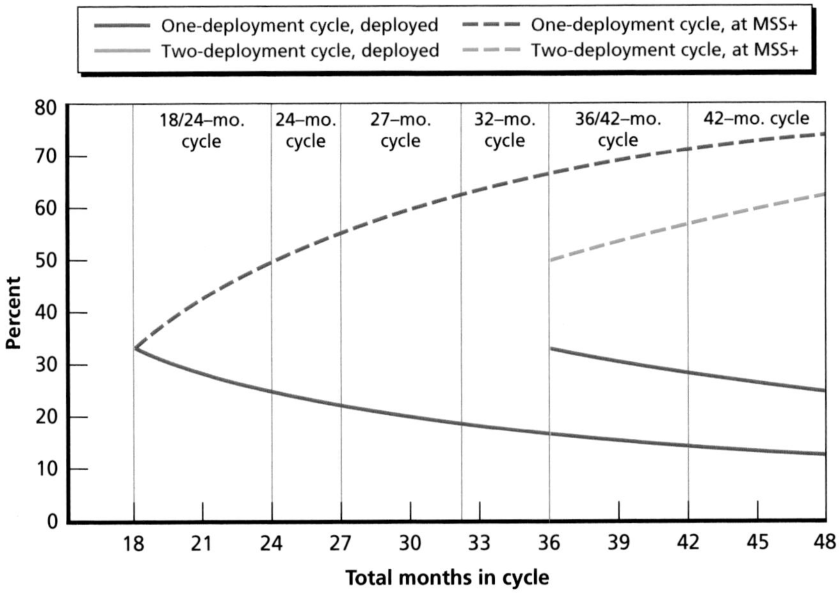

RAND *MG706-3.3*

Table 3.1
Summary Operational Measures for Various Cycles: Notional Carrier

	One Deployment				Two Deployments	
Cycle Length	**18/24-mo.**	**24-mo.**	**27-mo.**	**32-mo.**	**36/42-mo.**	**42-mo.**
Percentage of time deployed	31	25	22	19	31	29
Percentage of time at MCO-S/MCO-R status	15	30	36	46	39	44
Percentage of time at MSS status	18	15	14	11	10	9
Percentage of time in maintenance	36	30	28	24	20	18

Application of Alternative Cycles to the Carrier Fleet

The above results focus on deployments and deployability for a notional carrier. We now examine how cycles of differing length affect the readiness of the entire *Nimitz*-class fleet from fiscal year (FY) 2007 through FY 2025.[1]

Table 3.2 shows the proportion of time the fleet could meet the 6+1 fleet goal in the specified timeframe under the varying maintenance cycles we consider. Under the current 32-month cycle, the carrier fleet meets the 6+1 fleet goal 86 percent of the time. Under the previous 24-month cycle, the fleet meets the goal only 76 percent of the time. The two-deployment cycles meet the goal more often: 88 percent of the time for the 36/42-month cycle and 99 percent of the time for the 42-month cycle.

If availability goals were reduced to 5+1 or even 4+1, then the ability of each cycle to meet these new goals would increase. Given a 5+1 goal, three of the six cycles would meet availability goals more than 95

Table 3.2
FRP Metrics for Various Cycles: Carrier Fleet, FY 2007–2025

	One Deployment				Two Deployments	
Cycle Length	**18/24-mo.**	**24-mo.**	**27-mo.**	**32-mo.**	**36/42-mo.**	**42-mo.**
Percentage of fleet at 6+1	64	76	75	86	88	99
Percentage of fleet at 5+1	89	90	93	97	100	100
Percentage of fleet at 4+1	99	98	99	100	100	100

1 We assume the following: The *Enterprise* and the *Kitty Hawk*, carriers outside the *Nimitz* class that will retire in a few years, will remain on their current cycles. *Ford*-class ships will be on a 48-month, two-deployment cycle with one docking before the RCOH and another after, per information from NAVSEA 08. Future delivery and decommissioning will occur in accordance with the dates presented in Table 2.1, with PSAs for future carriers completed within one year of delivery to the fleet and the RCOH fixed at a ship's midlife point. The current DPIA3 period for the USS *Nimitz*, November 2010 to February 2012, is fixed. New cycles begin in June 2008 for any carrier not in a depot availability, with any carrier in a depot availability beginning the new cycle at the end of the availability.

percent of the time. Given a 4+1 goal, all six cycles would meet availability goals at least 98 percent of the time.

Figure 3.4 provides more detail on the ability of the 32-month cycle to meet the 6+1 fleet goal over time. It shows the average number of carriers that would be deployed, deployable within 30 days (i.e., at MCO readiness), and deployable within 90 days (i.e., at MSS readiness). The blue horizontal line drawn at 2.7 average carriers per year shows the threshold that must be deployed to meet current theater presence requirements. This includes the transit time required to provide an annual average of 2.3 in-theater carriers (represented by the red horizontal line).

Under the current 32-month cycle, average annual deployment requirements will not be met in most years. In particular, they will not be met from FY 2013 to FY 2016, a four-year period when there will be only ten carriers in the fleet. Between FY 2013 and FY 2015, the average number of deployed and readily deployable carriers would fall below the six needed for the 6+1 fleet goal. The USS *Gerald R. Ford,*

Figure 3.4
Current 32-Month Cycle Applied to the Fleet

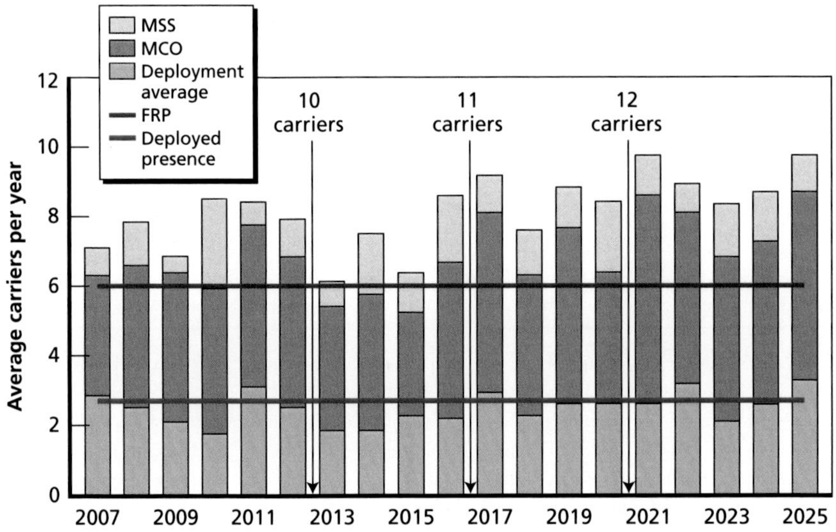

RAND *MG706-3.4*

though launching in FY 2015, will likely not be available to deploy until FY 2017. Similarly, CVN 79, though launching in FY 2019, will likely not be available to deploy until FY 2021. Data for the past four carriers (CVN 73 through CVN 76) indicates that it takes an average of 30 months from launch to the first deployment.

Figure 3.5 shows deployment metrics for the 18/24-month cycle. This cycle better meets presence requirements than does the 32-month cycle, doing so in all years but one. However, it falls short of meeting the 6+1 fleet goal in many years.

Figure 3.6 shows the deployment metrics for the 36/42-month, two-deployment cycle. Under this cycle, the fleet would be able to meet presence requirements, with an average annual number of 2.7 ships deployed in virtually every year between now and 2025. It would also meet the 6+1 fleet goal in all but a few years.

Figure 3.7 shows the deployment metrics for the 42-month, two-deployment cycle. Under this cycle, the fleet would meet the 2.7 average deployment goal in most years, but not in as many years as the

Figure 3.5
18/24-Month Cycle Applied to the Fleet

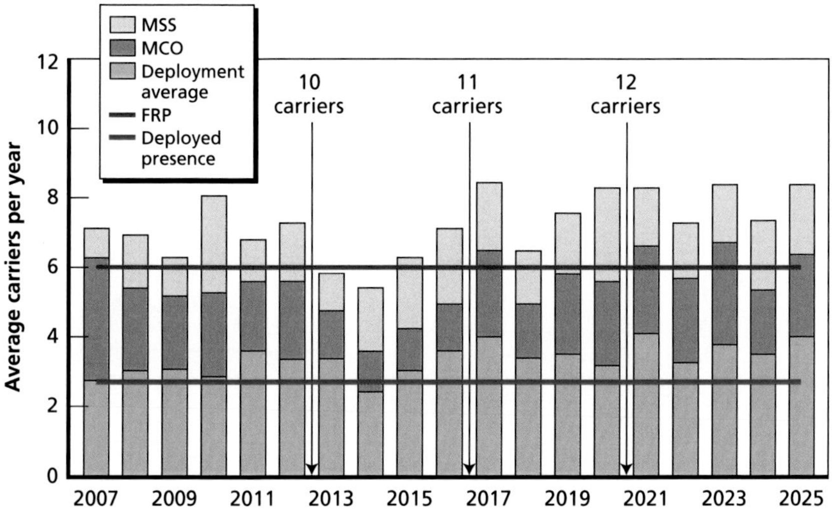

RAND *MG706-3.5*

Figure 3.6
36/42-Month Cycle Applied to the Fleet

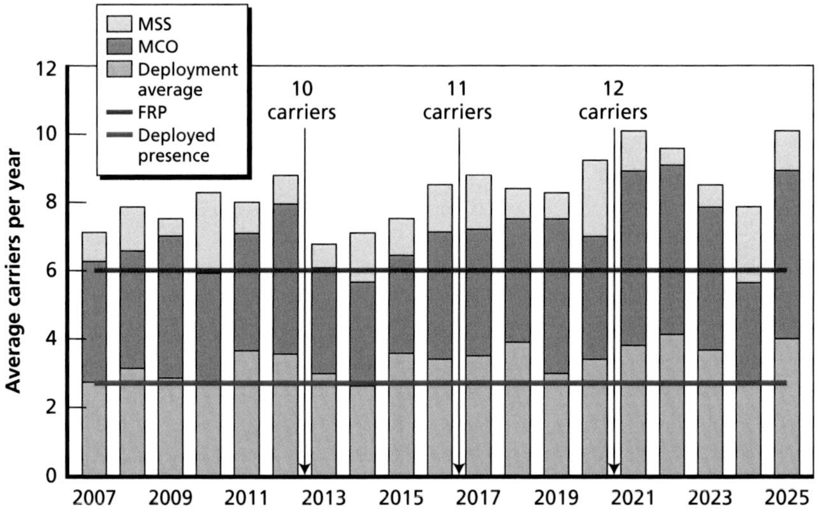

RAND *MG706-3.6*

Figure 3.7
42-Month Cycle Applied to the Fleet

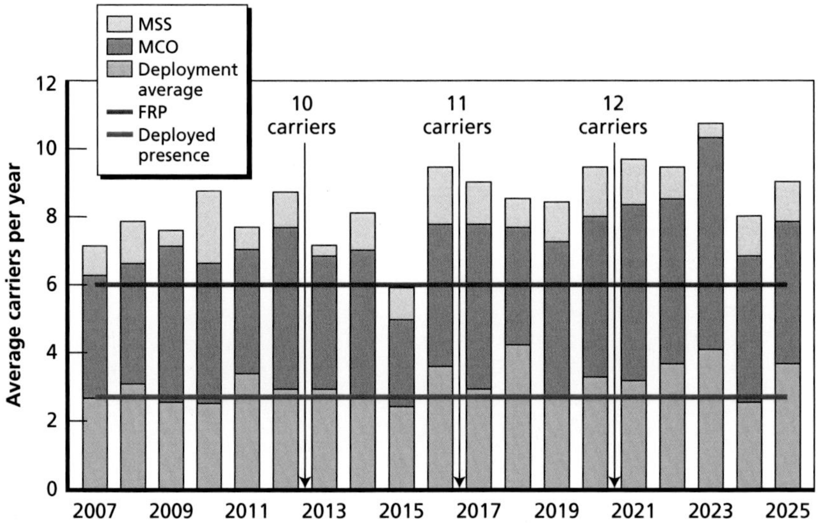

RAND *MG706-3.7*

36/42-month cycle. It would exceed the 6+1 fleet goal in virtually every year, but fall short of even a 5+1 goal in 2015.

Our analysis suggests that two-deployment cycles can help increase forward presence to the desired levels and maintain the overall readiness of the carrier fleet to meet contingencies and crises. We now turn to the impact of the different cycles on the depot maintenance industrial base and the cost of providing depot maintenance to the carrier fleet.

CHAPTER FOUR

The Impact of Different Cycles on the Maintenance Industrial Base

Our analysis in previous chapters describes the varied effects of each of the six cycles on the operational availability of the fleet. In this chapter, we examine the potential impact of each cycle on the maintenance industrial base and the costs of depot maintenance for carriers. We study the ways in which each cycle could affect the total maintenance workload over the life of a *Nimitz*-class carrier, how resulting workloads would be distributed over depot maintenance availabilities, and how workloads could affect the demand for labor at Norfolk Naval Shipyard (NNSY) and Puget Sound Naval Shipyard (PSNSY), the two Navy shipyards that accomplish depot availabilities for the aircraft carrier fleet.[1] Finally, we estimate the potential impact of varying cycles on depot maintenance costs.

Our availability work package estimates for the new cycles are initial, rough order-of-magnitude estimates based on several simplifying assumptions. As noted earlier, detailed engineering studies will be required to fully understand which maintenance tasks can be extended from a 32-month interval to 36- or 42-month intervals. Those tasks

---

[1] Northrop Grumman Newport News, a private company, builds all new aircraft carriers and performs the PSAs and RCOHs for *Nimitz*-class aircraft carriers. It also performs all depot availabilities for the USS *Enterprise* and occasionally performs PIAs or DPIAs for *Nimitz*-class carriers. The ship repair facility (SRF) in Yokosuka, Japan, supports the aircraft carrier that is part of the FDNF. The Japanese government provides some of the SRF's workforce, but PSNSY will perform all nuclear-related work when the USS *George Washington* joins the FDNF.

that cannot be extended may have to be accomplished within the longer cycles, perhaps in a CIA. Further analysis by organizations with deeper knowledge of carrier maintenance requirements, such as the CPA, would also be needed to better define and schedule the specific tasks and workloads for depot availabilities in longer cycles, particularly given limited shipyard resources.

Estimating the Magnitude of Depot Work Packages

The IMP for *Nimitz*-class carriers defined a 24-month cycle between depot availabilities with a 72-month cycle between docking availabilities. Each 72-month cycle included two PIAs followed by a DPIA. This PIA–PIA–DPIA cycle was repeated four times over the first half of a *Nimitz*-class carrier's life; the final DPIA was an RCOH (see Figure 2.4, above). The cycle was repeated following the RCOH.

Because maintenance requirements increase as a ship ages, the IMP defined three PIA and DPIA work packages. The PIA1 and DPIA1 work packages were set at 85 percent of the corresponding PIA2 and DPIA2 work packages, and the PIA3 and DPIA3 work packages were set at 115 percent of the PIA2 and DPIA2 packages. PIA1 and DPIA1 work packages occurred only before the RCOH; the first maintenance cycle after an RCOH for a *Nimitz*-class carrier had PIA2/DPIA2 work packages, followed by three sets of PIA3/DPIA3 work packages. Table 4.1 shows this sequence and the IMP work packages for the 24-month cycle.

The FRP extension of the 24-month cycle to 27 months resulted in the loss of a PIA3 availability both before and after the RCOH. This, in turn, resulted in approximately 400,000 fewer man-days of depot work over the life of a Nimitz-class carrier. It is unclear if these lost man-days were actually needed during the life of a Nimitz-class carrier or how, if at all, they would be reclaimed if they were needed. Since the Nimitz-class carriers were in fact operating on a 27-month cycle (see Figure 2.2, above), it is possible that the original estimate of workloads under the IMP were slightly inflated. Or, the depot system may have actually accomplished more work during the PIAs and DPIAs than

Table 4.1
Notional Work Packages for 24-, 27-, and 32-Month Cycles (thousands of man-days)

	24-mo. Cycle	27-mo. Cycle	32-mo. Cycle	
	Workload	Workload	Workload	CIA
PSA/SRA	71.0	71.0	71.0	N/A
PIA 1	146.2	146.2	146.2	18.0
PIA 1	146.2	146.2	146.2	18.0
DPIA 1	255.8	255.8	308.9	18.0
PIA 2	173.8	173.8	173.8	21.2
PIA 2	173.8	173.8	173.8	21.2
DPIA 2	308.9	308.9	356.6	21.2
PIA 3	201.4	201.4	201.4	24.4
PIA 3	201.4	201.4	201.4	24.4
DPIA 3	356.6	356.6	N/A	N/A
PIA 3	201.4	201.4	N/A	N/A
PIA 3	201.4	N/A	N/A	N/A
RCOH	3,200.0	3,200.0	3,200.0	24.4
PSA/SRA	71.0	71.0	71.0	N/A
PIA 2	173.8	173.8	173.8	21.2
PIA 2	173.8	173.8	173.8	21.2
DPIA 2	308.9	308.9	356.6	24.4
PIA 3	201.4	201.4	201.4	24.4
PIA 3	201.4	201.4	201.4	24.4
DPIA 3	356.6	356.6	356.6	24.4
PIA 3	201.4	201.4	201.4	24.4
PIA 3	201.4	201.4	201.4	24.4
DPIA 3	356.6	356.6	N/A	N/A
PIA 3	201.4	201.4	N/A	N/A
PIA 3	201.4	N/A	N/A	N/A
Total	8,287.0	7,884.2	7,296.3 (workload and CIA)	

Table 4.1—Continued

SOURCES: For the 24-month cycle, Department of the Navy, "Representative Intervals, Durations, Maintenance Cycles, and Repair Mandays for Depot Level Maintenance Availabilities of U.S. Navy Ships," May 1, 2002. For the 27-month cycle, Department of the Navy, "Representative Intervals, Durations, Maintenance Cycles, and Repair Mandays for Depot Level Maintenance Availabilities of U.S. Navy Ships," June 13, 2002. For the 32-month cycle, CPA, "USS Nimitz (CVN68)-Class Aircraft Carrier 32-Month Maintenance Cycle Notional Analysis," unpublished analysis, April 23, 2006, and Department of the Navy, 2007c.

was suggested by the notional work packages in OPNAV Notice 4700 of May 1, 2002. It is also possible that not enough time had accumulated under the 24-month cycle to make an informed judgment of the appropriate size for the work packages.

Table 4.1 shows the sequence of depot work packages for the 27-month cycle. Work packages are shown in sequential order for each cycle and may not correspond across rows of the table.

Compared to the 24-month cycle, the recent lengthening of the cycle to 32 months has resulted in one fewer DPIA3 availability and two fewer PIA3 availabilities both before and after the RCOH. Because the 32-month cycle extends the docking interval to 96 months, the first two DPIAs were set as the former DPIA2 and DPIA3; that is, the 32-month cycle has no DPIA1. Also, one CIA is scheduled during the years when there is no depot availability planned within each 32-month cycle. As with the PIAs and DPIAs, the size of the CIA work packages grows as the ship ages.[2] These changes partially compensate for the man-days lost from the reduced number of PIAs and DPIAs.[3] Table 4.1 also shows CPA's original estimate of the 32-month availability work packages.

Depot availabilities feature several types of work, including:

---

[2] The number and the size of the CIAs within each cycle have not yet been fully defined. In late 2007, the one fairly large CIA per cycle was reconfigured to form two, smaller CIAs in each cycle. The total amount of work across all the CIAs remained about the same.

[3] An availability consists of some services work that supports the direct labor activities. Examples include riggers, crane operators, and availability management. Therefore, fewer availabilities will result in less support work over the life of a carrier. This means that not all of the shortfall in work resulting from fewer availabilities in the longer cycles must be recovered in the CIAs.

- assessments—tests and inspections to determine if equipment or subsystem repair or replacement is required
- tank work—preservation of tanks and voids in the ship that requires opening and inspecting the tank or void, followed by blasting and repainting as required
- nuclear work—tests, inspections, and repair of nuclear components of the propulsion system.
- repairs—repair or replacement of failed components or pieces of equipment
- fleet alterations (formerly known as "D" ship alterations)—modernization work that involves permanent changes to a ship but does not affect the military characteristics of a ship. These alterations are funded by the Fleet or Type Commander and typically consist of reliability, maintainability, safety, and quality-of-life items.
- program alterations (formerly known as "K" ship alterations)—modernization work that involves permanent changes to the ship to provide a military characteristic, upgrade existing systems, or provide additional capability not previously held by the ship. The System Commands or the Program Executive Officer for the class of ship funds program alterations.
- services—the support provided to accomplish the above categories of work. This includes activities such as staging, rigging, and crane operations.

On average, services and repairs constitute most work within a PIA or DPIA, with nuclear work and program alterations combined making up about a third. Figure 4.1 shows a typical distribution of categories of work in a depot availability.

We also estimated the total workload in PIA, DPIA, and CIA work packages over the life of a carrier for the 18/24-, 36/42-, and 42-month cycles. We use the 32-month cycle to bracket our estimates at the lower bound and the 27-month cycle to bracket them at the upper bound. We start by subtracting the workload for the RCOH and the two PSA/SRAs from the totals in Table 4.1. We then assume that

Figure 4.1
Breakout of Typical Carrier Depot Work Package Content

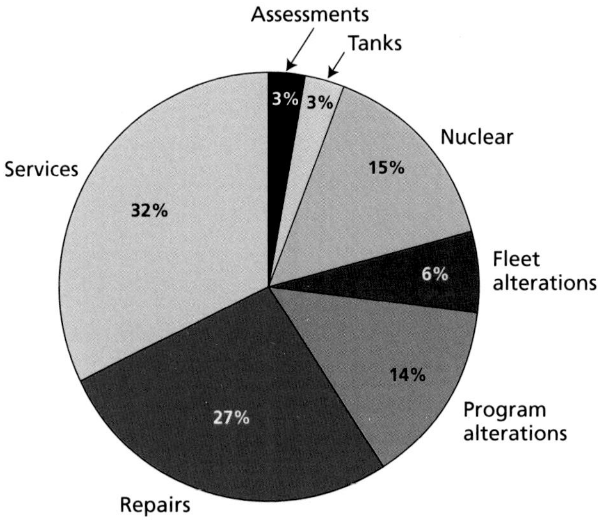

SOURCE: CPA.
NOTE: This breakout represents the combined depot work package content averages of 14 PIAs and 2 DPIAs.
RAND *MG706-4.1*

- the total work for assessments, tanks, nuclear, fleet alterations, and program alterations remains constant over the life of the carrier regardless of the maintenance cycle
- repair work varies directly with the number of days that the carrier is underway for training, sustainment, or deployments
- services are 47 percent (32 percent divided by 68 percent) of the sum of the other categories of work.

We estimate the total number of underway days for each cycle by assuming that a 6-month deployment has 25 underway days per month and that non-deployed time spent in training or sustainment involves 24 steaming days per quarter (or eight days per month). Of course, there are no steaming days when the carrier is in a maintenance availability. Also, we do not include the last deployment before a carrier is decommissioned. Table 4.2 shows the resulting calculations for the

Table 4.2
Deployments and Underway Days During a Carrier's Life

	One Deployment				Two Deployments	
	18/24-mo. Cycle	24-mo. Cycle	27-mo. Cycle	32-mo. Cycle	36/42-mo. Cycle	42-mo. Cycle
Number of deployments before final deployment	29	23	21	17	31	27
Training/sustainment underway days[a]	1,472	2,018	2,328	2,608	2,240	2,512
Deployment underway days[b]	4,350	3,450	3,150	2,550	4,650	4,050
Total underway days	5,822	5,466	5,478	5,158	6,890	6,562

[a] Underway days during training/sustainment = 24 days per quarter.

[b] Underway days during deployment = 25 days per month.

number of deployments and underway days for deployment and training or sustainment for the six cycles we examine.

Similarly, Table 4.3 shows the resulting calculations for our estimates of the total PIA and DPIA workload for the 18/24-, 36/42-, and 42-month cycles.

We next distribute the total PIA, DPIA, and CIA workloads over the availabilities of the 18/24-, 36/42-, and 42-month cycles. To do this, we assume that

- based on the 27 and 32-month cycles, approximately 47 percent of the total workload is accomplished in the first half of a carrier's life (with the remaining 53 percent accomplished after the midlife RCOH).
- the first set of availabilities is 85 percent of the second set and the third set of availabilities is 115 percent of the second set.
- the first half of a carrier's life has a DPIA2 and DPIA3 for the 36/42-month cycle. (There is only a single DPIA2 in the first half of the 42-month cycle.)

Table 4.3
Estimates of PIA/DPIA Workloads for 18/24-, 36/42-, and 42-Month Cycles

	27-mo. Cycle	32-mo. Cycle	18/24-mo. Cycle		36/42-mo. Cycle		42-mo. Cycle	
			Lower Bound	Upper Bound	Lower Bound	Upper Bound	Lower Bound	Upper Bound
Assessments, tank and nuclear work, alterations	1,862[a]	1,621[a]	1,621	1,862	1,621	1,862	1,621	1,862
Repair	1,226[b]	1,068[b]	1,205	1,303	1,427	1,542	1,359	1,469
Underway days	5,478	5,158	5,822 (lower and upper)		6,890 (lower and upper)		6,562 (lower and upper)	
Services[c]	1,454	1,265	1,328	1,488	1,433	1,600	1,401	1,566
Total	4,542	3,954	4,154	4,653	4,481	5,004	4,381	4,897

NOTES: All units are in thousands of man-days except the underway days. Lower bounds are based on a 32-month cycle; upper bounds are based on a 27-month cycle.

[a] Forty-one percent of total man-days.

[b] Twenty-seven percent of total man-days.

[c] Thirty-two percent of total man-days.

- the availabilities after the RCOH are PIA2s, DPIA2s, and CIA2s. (The 42-month cycle only has one DPIA3 in the second half of the carrier's life.)
- there are two CIAs per cycle for the 36/42- and 42-month cycles.

Table 4.4 shows the resulting estimated work package for the 18/24-, 36/42-, and 42-month cycles, as well as for the 24-, 27-, and 32-month cycles.

The DPIA work packages for the 18/24-month cycle are very similar in size to those for the other one-deployment cycles. Yet because there are 24 PIAs over the lifetime of a carrier on the 18/24-month cycle, compared to 12 PIAs for one on the 32-month cycle and 14 for one on the 27-month cycle, the size of the PIA work packages for the 18/24-month cycle (shaded in green in Table 4.4) are much smaller. They are 100,000–150,000 man-days for the 18/24-month cycles, compared to 146,000–201,000 man-days for the 27-month cycles. Because a Navy depot can efficiently execute approximately 30,000 man-days of work per month during a carrier availability,[4] the PIAs for the 18/24-month cycle could be shorter, possibly four rather than six months. This would increase the time a carrier is deployable within 90 days (i.e., in MCO-S or MCO-R status) by approximately 10 percent. With shorter PIA durations, the 18/24-month cycle begins to look like the old EOC maintenance cycle.

For the two-deployment cycles, a smaller number of PIAs and DPIAs combined with roughly comparable total work requirements result in much larger individual PIA and DPIA work packages. Under the original 24-month cycle, there were sixteen 6-month PIAs and six 10-½-month DPIAs over the life of a *Nimitz*-class carrier. The 27-month cycle eliminated two PIAs. The 36/42-month cycle, with its ten PIAs, four DPIAs, and 32 one-month CIAs, has still fewer PIAs. The 42-month cycle has ten PIAs, two DPIAs, and 28 CIAs.

---

[4] Shipyard officials estimate that approximately 1,500 personnel can work efficiently on a carrier at a given time.

Table 4.4
Work Package Estimates for Different Cycles (thousands of man-days)

	One Deployment				Two Deployments	
Work Package	**18/24-mo. Cycle**	**24-mo. Cycle**	**27-mo. Cycle**	**32-mo. Cycle**	**36/42-mo. Cycle**	**42-mo. Cycle**
PIA1	100–115	146	146	146	200–220	255–280
PIA2	120–135	174	174	174	230–250	290–320
PIA3	135–150	201	201	201	265–290	340–375
DPIA1	250–275	256	256	N/A	N/A	N/A
DPIA2	280–320	309	309	309	330–360	365–400
DPIA3	320–370	357	357	357	380–420	410–450
CIA1	N/A	N/A	N/A	18	15–20[a]	15–20[a]
CIA2	N/A	N/A	N/A	21	20–25[a]	20–25[a]
CIA3	N/A	N/A	N/A	24	25–30[a]	25–30[a]
Total	4,154–4,653	4,945	4,542	3,954	4,481–5,004	4,381–4,897

[a] Two CIAs per cycle.

Given that a Navy depot can manage approximately 30,000 man-days of work per month during a PIA or DPIA, it is not clear whether the resulting larger work packages for the longer cycles can be managed. Most of the two-deployment work packages (shaded in red in Table 4.4) would exceed 30,000 man-days per month. Some would require more than twice as many workdays as a shipyard could accommodate in six months. Accordingly, the duration of the two-deployment work packages may exceed the six months allocated for a PIA and the 10-½ months allocated for a DPIA.

Continuing maintenance availabilities may also pose problems for some cycles. Due to the inefficiencies of performing depot-level work at the operating base and the limitations on the length of the workday due to the crew living on the ship, the CPA estimates that, at most, 15,000 man-days can be executed during a month-long CIA period, or about half of what may be required for some later CIA periods.

Should work packages exceed the time allocated for them, there would be three additional adverse effects. First, the carrier crew members that work with the nuclear propulsion system may require additional recertification and retraining, thereby lengthening the required training time after the depot availability. Second, the longer time for the depot availabilities and the increased training would reduce the amount of time a carrier is at surge-ready status. Third, planners believe that longer availabilities become still more unpredictable in length and increase the possibility that work on a ship will not be completed in time. If the duration of the PIAs and DPIAs were to grow for the two-deployment cycles, there would be a reduction in the percentage of time that a carrier is at MCO-S or higher readiness status.

This suggests that gaining more forward presence through two-deployment cycles involves tradeoffs between maintenance duration and surge capability. While two deployments in a cycle can yield more forward presence, their maintenance periods then result in less 6+1 surge readiness.

Impact on the Depots

We apply these notional workloads to the current and planned carrier fleet, using the rules described in Chapter Three for calculating the operational availability measures. We then assess the overall impact on the workload projections for NNSY and PSNSY, the two public shipyards that perform most *Nimitz*-class PIAs and DPIAs.

Norfolk Naval Shipyard

The estimated FY 2007–2017 workload at NNSY for the 18/24-month cycle, when applied to the current and future fleet, is shown in Figure 4.2. Figure 4.3 displays the same information for the 42-month cycle.[5] These two cycles represent the extremes in terms of duration of the

Figure 4.2
Total Workload at NNSY: 18/24-Month Cycle

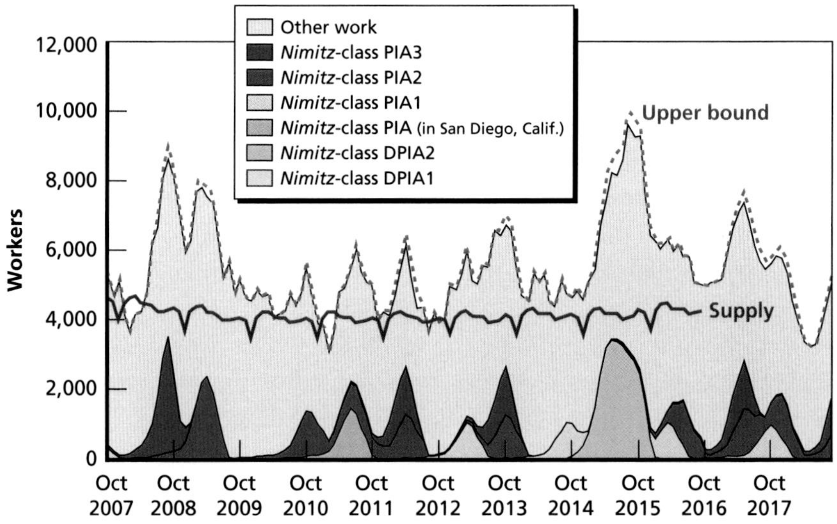

NOTE: Supply data provided by the workload allocation and resource report (WARR) file used for NNSY did not extend past October 2016.
RAND *MG706-4.2*

---

[5] The appendix supplies the individual workload graphs at NNSY and PSNSY for each cycle.

cycle and the size of the workload packages. The 18/24-month cycle has small work packages that occur frequently. The 42-month cycle has large work packages that occur infrequently. The carrier work is shown in color. We also show other, non-carrier work to project the total workload at the shipyard.[6] The dotted lines represent the upper estimates of the PIA and DPIA workloads. The red line represents the current and projected supply of full-time workers, based on data from the individual shipyard's WARR files. Each WARR file contains data

Figure 4.3
Total Workload at NNSY: 42-Month Cycle

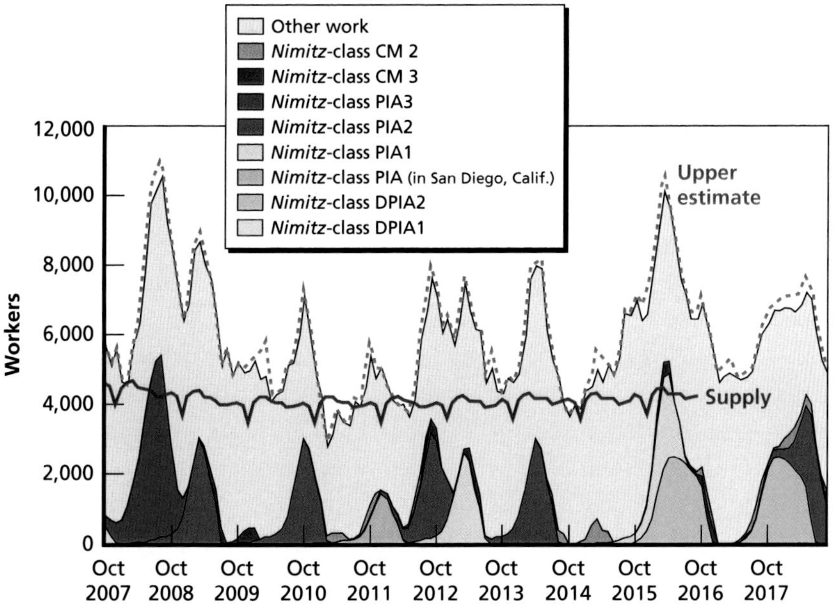

NOTE: Supply data provided by the WARR file used for NNSY did not extend past October 2016. CM = continuous maintenance.
RAND *MG706-4.3*

---

6 These workloads are based on a March 2006 Common Overhaul File (COF) supplied by NAVSEA 04, the Logistics, Maintenance, and Industrial Operations Directorate. This file lists all availabilities being performed by the four public shipyards.

on the current and projected workforce. We used a February 2006 WARR file provided by NNSY to generate this figure.

Under the 18/24-month cycle, carrier workload at NNSY causes the total demand to exceed 8,000 workers (more than twice the number of workers in the shipyard) twice in the next decade; demand exceeds 6,000 workers several other times over the same period. Under the 42-month cycle, carrier workload causes the total demand to exceed 10,000 workers in at least two different periods. More importantly, all the cycles would generate workload estimates that almost always exceed the workforce supply line, often by substantial amounts. NNSY handles overload situations with overtime, the use of subcontracted workers, and the borrow-and-loan program that facilitates the sharing of workers between public shipyards. Extensive use of these stopgap measures could raise additional cost issues.

The 42-month cycle results in several gaps in carrier work at the shipyard. These gaps, which last from several months to over a year, may lead to the loss of expertise in performing carrier maintenance work, especially for tasks that are not performed during every carrier availability. This negative learning may in turn result in an increase in the hours needed to accomplish certain maintenance tasks. From a shipyard worker efficiency perspective, it is definitely desirable to have some level of continuous carrier work at the shipyard. The longer, two-deployment cycles may preclude this option.

Table 4.5 displays both the length of periods in which workload demand exceeds worker supply as well as when the supply of workers exceeds the demand at NNSY. For each cycle, the table shows the total workload above and below supply, the number of months that workload is projected to be above or below supply, and the average amount per month by which workload will be above or below supply. The values for the current 32-month cycle are highlighted.

These averages help compare cycles and their workload demands more succinctly, albeit at the expense of obscuring some peak problem periods evident in Figure 4.2. The 32-month cycle results in the least total excess of monthly demand over supply. Nevertheless, there appears to be minimal difference between the total excess in it and that for the lower estimates of workload for the 24-, 36/42-, and 42-month

Table 4.5
Comparison of Different Cycles: NNSY

Cycle	Demand Greater Than Supply			Supply Greater Than Demand		
	Total workload[a]	Number of months	Average per month[a]	Total workload[a]	Number of months	Average per month[a]
18/24-mo. cycle	173	98	1.60	4	10	0.04
24-mo. cycle	192	95	1.80	6	13	0.06
27-mo. cycle	184	99	1.70	4	9	0.04
32-mo. cycle	167	96	1.50	5	12	0.04
36/42-mo. cycle	175–207	97–102	1.60–1.90	3–4	6–11	0.03–0.05
42-mo. cycle	184–219	92–97	1.70–2.00	5–7	11–16	0.04–0.07

[a] In thousands of man-days.

cycles. The differences are certainly within the accuracy of the workload estimates.

We estimate little difference in depot maintenance cost between the current 32-month cycle and the depot cost for the 18/24-month cycle or for the lower workload estimates of the two-deployment cycles. If the PIA and DPIA work packages for the two-deployment cycles increase to the level of the upper estimates, depot maintenance costs could increase by as much as $22 million–28 million dollars (40,000 to 50,000 man-days × $550 per man-day),[7] or less than $2 million annually.

Puget Sound Naval Shipyard

Figure 4.4 shows the estimated FY 2007–2017 carrier and total workload at PSNSY for the 18/24-month cycle, along with the estimated supply of workers. The other, non-carrier work is also shown to project

[7] The man-day rate is based on the average private sector man-day rates in the Midatlantic and West Coast regions.

Figure 4.4
Total Workload at PSNSY: 18/24-Month Cycle

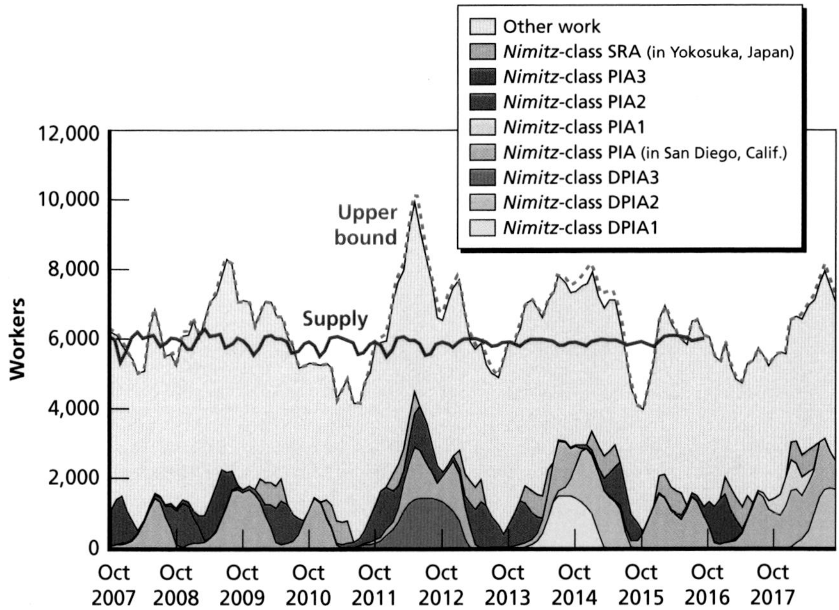

NOTE: Supply data provided by the WARR file used for PSNSY did not extend past October 2016.
RAND *MG706-4.4*

the total workload at the shipyard. Figure 4.5 displays the same information for the 42-month cycle. The sources and format are otherwise similar to Figures 4.2 and 4.3. We used a June 2006 WARR provided by PSNSY to generate Figure 4.5.

Under the 18/24-month cycle, carrier workload results in one period of demand for more than 10,000 workers, or more than 1.5 times the supply of workers in the shipyard; several times, demand would briefly require more than 8,000 workers. Under the 42-month cycle, demands for carrier maintenance also result in a period in which workload demand briefly exceeds 10,000 workers, another in which it approaches it, and several in which it approaches or exceeds 8,000 workers. Overall, PSNSY differs from NNSY in having more instances of worker supply exceeding workload demand. Table 4.6 summarizes

Figure 4.5
Total Workload at PSNSY: 42-Month Cycle

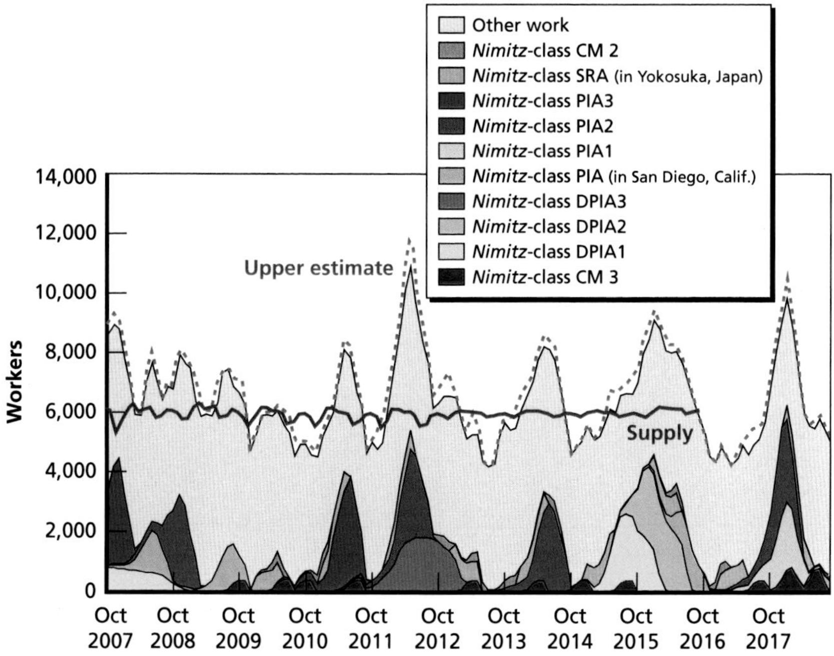

NOTE: Supply data provided by the WARR file used for PSNSY did not extend past October 2016.
RAND *MG706-4.5*

the length of periods in which worker supply does not match workload demand. The table obscures the extent of difficulties in peak problem periods.

Table 4.6 suggests that lengthening the maintenance cycle to 32 months has helped PSNSY reduce the number of months in which workload demand exceeds worker supply. At the same time, this shift has also resulted in an increase in the number of months in which worker supply exceeds demand. An 18/24-month cycle would result in more months in which demand exceeds supply as well as a greater total excess workload during those months, but it would also result in fewer months in which supply exceeds demand. The effects on PSNSY of implementing a two-deployment cycle are not clear. Should the lower

Table 4.6
Comparison of Different Cycles: PSNSY

Cycle	Demand Greater Than Supply			Supply Greater Than Demand		
	Total Workload[a]	Number of Months	Average per Month[a]	Total Workload[a]	Number of Months	Average per Month[a]
18/24-mo. cycle	64	62	0.6	38	46	0.4
24-mo. cycle	78	45	0.7	65	63	0.6
27-mo. cycle	57	43	0.5	63	65	0.6
32-mo. cycle	38	30	0.4	98	78	0.4
36/42-mo. cycle	16–86	24–71	0.2–0.8	29–93	37–84	0.8–0.9
42-mo. cycle	31–121	36–72	0.3–1.1	20–64	36–72	0.2–0.6

[a] In thousands of man-days.

estimates of the PIA and DPIA work packages prove accurate, then there would still be excess workload here; if the higher estimates prove accurate, then there would be even more periods of excess workload. Yet even under the higher estimates for the 18/24-month cycle, there would be fewer months in which labor supply exceeds demand than under the current 32-month cycle.

Altogether, our analysis suggests that various cycles will have mixed effects on the workloads at the public shipyards. The longer, two-deployment cycles may result in a slight annual increase in cost at NNSY (when compared to the current 32-month cycle). The one-deployment cycle may complicate workforce management at PSNSY when workload demand exceeds the supply of workers, but will it alleviate the situation when supply exceeds demand. Similarly, the two-deployment cycles may cause workforce management problems when demand exceeds supply; but these problems may recede when supply exceeds demand.

CHAPTER FIVE

Findings and Recommendations

The depot maintenance program for *Nimitz*-class aircraft carriers has changed several times since the USS *Nimitz* was commissioned in 1975. After the EOC, the *Nimitz* class was transitioned to the IMP in 1994. Like the EOC, the IMP used a 24-month schedule for depot availabilities. However, the IMP level-loaded the workloads across those availabilities to improve the materiel condition of the carriers and to avoid the workload and funding spikes of the EOC. The FRP, instituted in 2003, modified the training and readiness of the carrier fleet to increase its ability to respond to emerging crises. It also increased the *Nimitz*-class depot maintenance cycle to 27 months. Most recently, the cycle has been lengthened to 32 months.

Increasing the length of the depot maintenance cycle has had several effects. Longer cycle lengths that retain depot availabilities of similar duration to those currently in place have decreased the proportion of time a carrier is in maintenance, thereby increasing the ability of the fleet to meet the 6+1 surge requirements of the FRP. Yet one-deployment cycles in this increased cycle length environment mean that the overall proportion of time that a carrier is deployed has decreased. Given the looming decrease in the carrier force, it has become difficult for operational planners to meet increasing demands for carrier presence in various theaters of operation.

The Commander, United States Fleet Forces Command, seeks to meet the challenge of providing carriers where and when they are needed. The needs are being met by adjusting personnel tempo policies for deployed ships. Deployments may be longer or shorter than six months, and carriers may redeploy in shorter timeframes than before. Our analysis offers options for increasing carrier forward presence by retaining deployment policies while examining the impact of longer or shorter depot maintenance cycles.

We examined shorter, one-deployment cycles and longer, two-deployment cycles. The alternatives we analyzed offer advantages and disadvantages compared to the current 32-month, one-deployment cycle. Figure 5.1 summarizes some of the advantages and disadvantages for three cycles we considered, assuming deployments of six months, PIAs of six months, and DPIAs of 10-½ months.

In comparison to the 32-month cycle, an 18/24-month cycle would increase the proportion of time a carrier is deployed (from 19 to 31 percent) and allow operational planners to meet an annual 2.7 carrier presence requirement over the next decade. Yet it would also increase the proportion of time a carrier is in maintenance from 24

Figure 5.1
Summary Operational Measures for 18/24-, 32-, and 42-Month Cycles (Over the Life of a Notional Carrier)

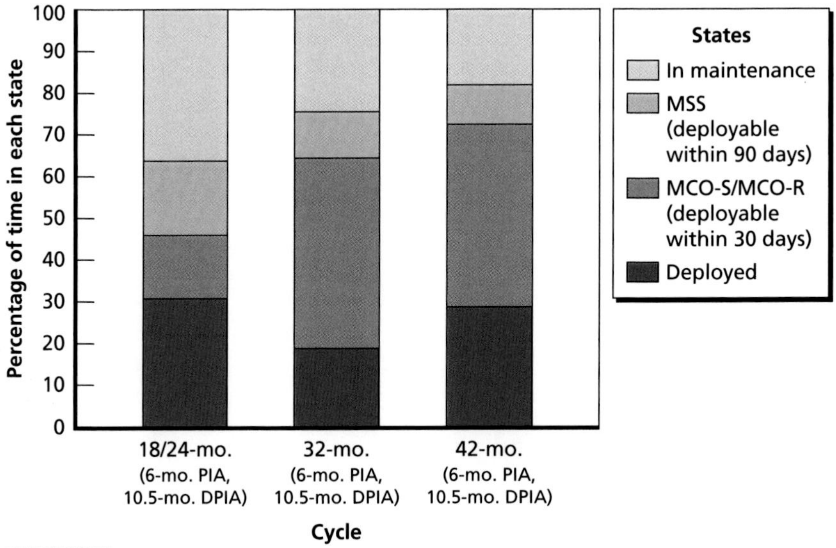

RAND *MG706-5.1*

to 36 percent. As a result, it would decrease the proportion of time a carrier could deploy in less than 90 days from 57 to 33 percent. This in turn would reduce the probability that the carrier could help meet the 6+1 fleet goal from 86 percent to 64 percent. Maintenance for the 18/24-month cycle is technically feasible, and the cycle would result in more frequent and uniform distribution of carrier-related work to NNSY and PSNSY. Our workload estimates suggest that spreading the carrier work over more PIAs and DPIAs would lead to smaller work packages. This in turn could lead to shorter PIAs and DPIAs and increase the proportion of a time a carrier could support the 6+1 fleet goal beyond what our assumptions indicate. We return to this possibility below.

Compared to the current 32-month cycle, the 42-month cycle would increase the proportion of time a carrier is deployed from 19 to 29 percent and retain similar surge capabilities. The higher proportion of time a carrier would be deployed under the 42-month cycle would better help operational planners to meet an annual 2.7 carrier forward-presence requirement over the next decade. Coupled with surge capabilities similar to those of the 24-month cycle, it would also increase the probability of meeting the 6+1 fleet goal to 99 percent.

Maintenance for a 42-month cycle may be more difficult to manage. Engineering studies would be required to assess these possibilities. The 42-month cycle would permit only a single docking before and after the midlife RCOH and a total of 12 depot availabilities over the life of a carrier (compared to 16 for the 32-month cycle). Even if the carrier depot workload were to remain unchanged with the additional underway days in the two-deployment cycle, the fewer opportunities for depot maintenance would lead to larger work packages. Our workload estimates suggest that the PIA, DPIA, and CIA work packages could grow to the point where they would not be executable in the durations we assumed. Also, the longer cycles would result in periods of several months to a year during which there would be no carrier work at a shipyard. This could lead to a loss of learning and an increased man-days requirement for maintenance tasks that are performed infrequently.

Figure 5.2
The Impact of Different PIA Durations on the Operational Availability of a Notional Carrier

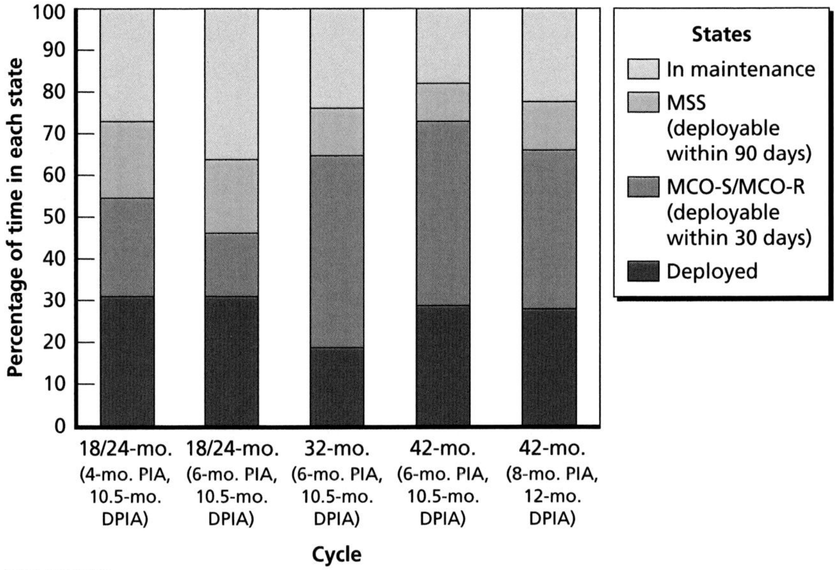

RAND *MG706-5.2*

Figure 5.2 shows the impact on various operational measures if the PIAs for the 18/24-month cycle were shortened to four months and the PIAs for the 42-month cycle were lengthened to eight months. The shorter-duration PIAs would increase the proportion of time a carrier is deployable within 90 days but not deployed (i.e., in MSS, MCO-S, or MCO-R status) from 33 to 42 percent. Four-month PIAs for the 18/24-month cycle would result in a depot maintenance program very similar to the old EOC program. Longer-duration PIAs for the 42-month cycle would decrease the proportion of time a carrier is deployable within 90 days but not deployed from 53 to 50 percent.

Our analysis suggests that shortening the one-deployment cycle from 32 months to 18/24 months would increase the forward presence of the carrier fleet but reduce its ability to meet the 6+1 fleet goal. Longer two-deployment cycles would increase forward presence while sustaining high levels of readiness. If the current plans of the Commander, United States Fleet Forces Command, to increase presence by modifying deployment lengths and turnaround times adversely affect retention and recruitment, then engineering studies should be funded to determine the technical feasibility of longer, two-deployment cycles and their potential impact on PIA, DPIA, and CIA work packages. Indeed, the fleet has experienced difficulties in meeting the 6+1 fleet goal even in a 32-month cycle, which suggests that such studies should be conducted now. This will allow the Navy to explore all options for improving the forward presence and deployability of its carriers.

APPENDIX

Workload Graphs for the Norfolk and Puget Sound Naval Shipyards

This appendix provides the various total workload graphs under the various carrier maintenance cycle options for NNSY and PSNSY.

Figure A.1
Total Workload for 18/24-Month Carrier Cycle—NNSY

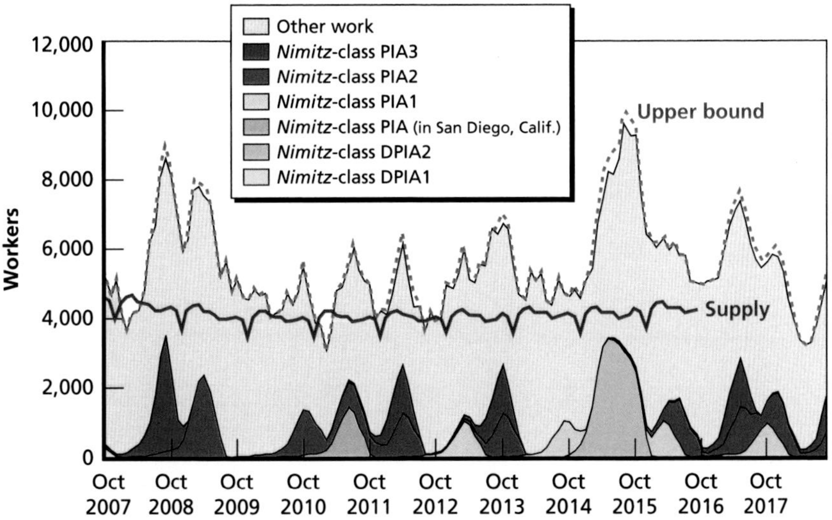

NOTE: Supply data provided by the WARR file used for NNSY did not extend past October 2016.

RAND *MG706-A.1*

Figure A.2
Total Workload for 24-Month Carrier Cycle—NNSY

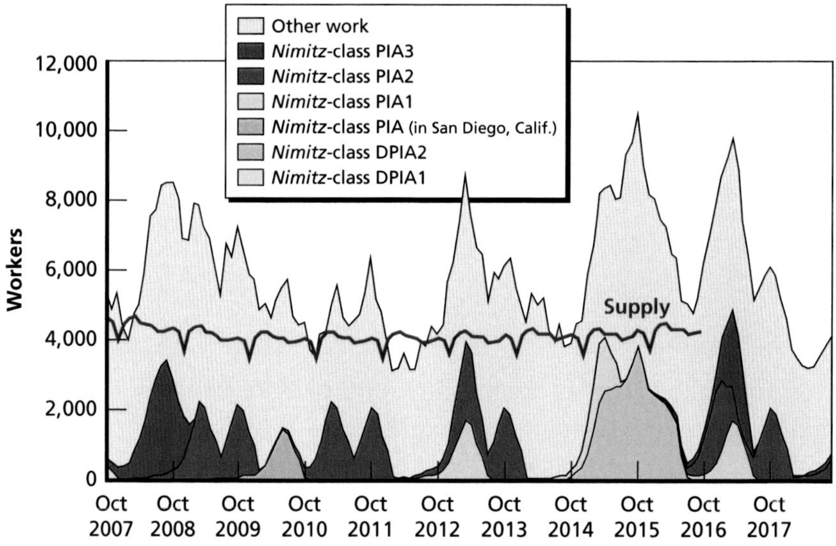

NOTE: Supply data provided by the WARR file used for NNSY did not extend past October 2016.

RAND *MG706-A.2*

Figure A.3
Total Workload for 27-Month Carrier Cycle—NNSY

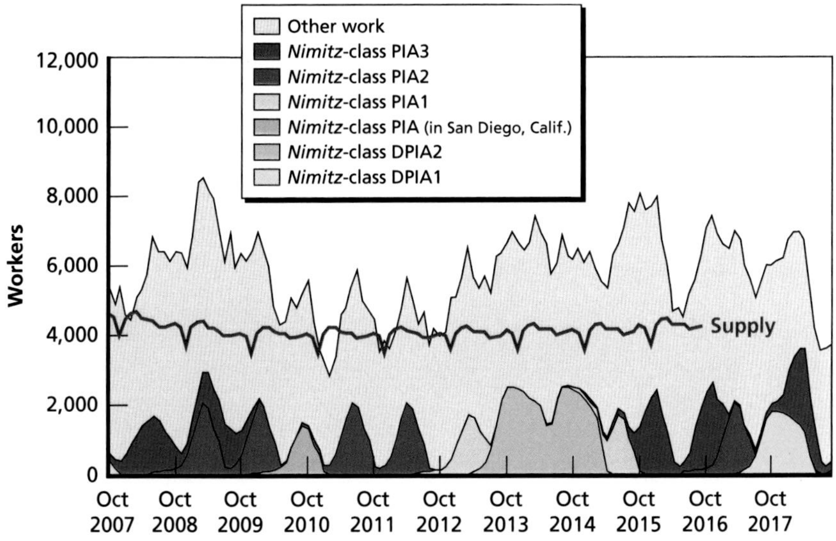

NOTE: Supply data provided by the WARR file used for NNSY did not extend past October 2016.

RAND *MG706-A.3*

Figure A.4
Total Workload for 32-Month Carrier Cycle—NNSY

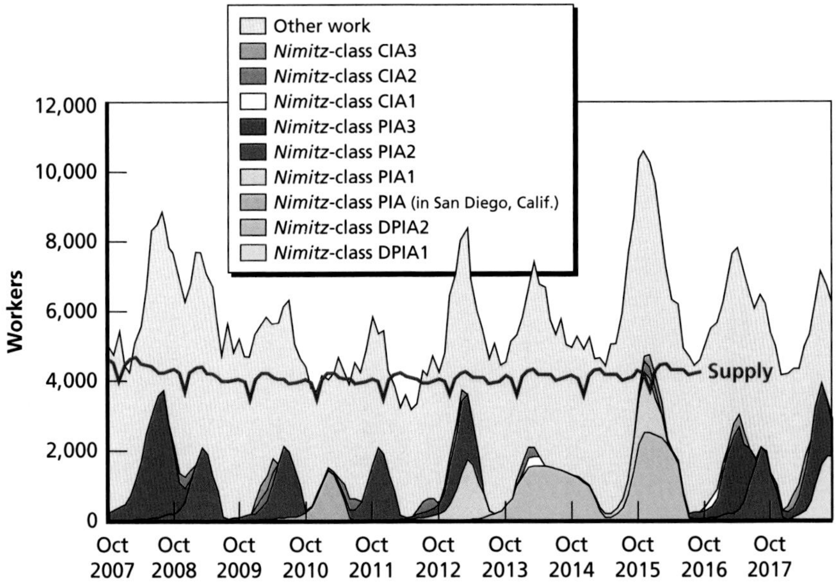

NOTE: Supply data provided by the WARR file used for NNSY did not extend past October 2016.

RAND *MG706-A.4*

Figure A.5
Total Workload for 36/42-Month Carrier Cycle—NNSY

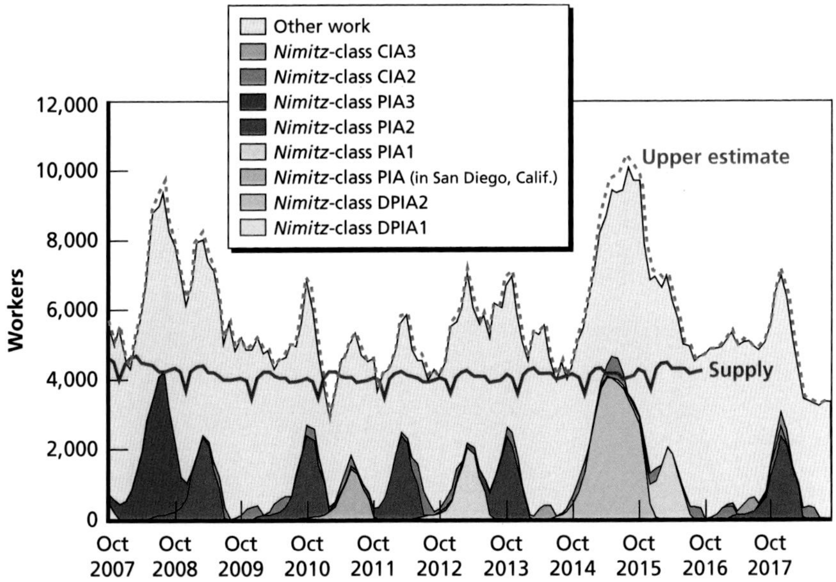

NOTE: Supply data provided by the WARR file used for NNSY did not extend past October 2016.

RAND *MG706-A.5*

Figure A.6
Total Workload for 42-Month Carrier Cycle—NNSY

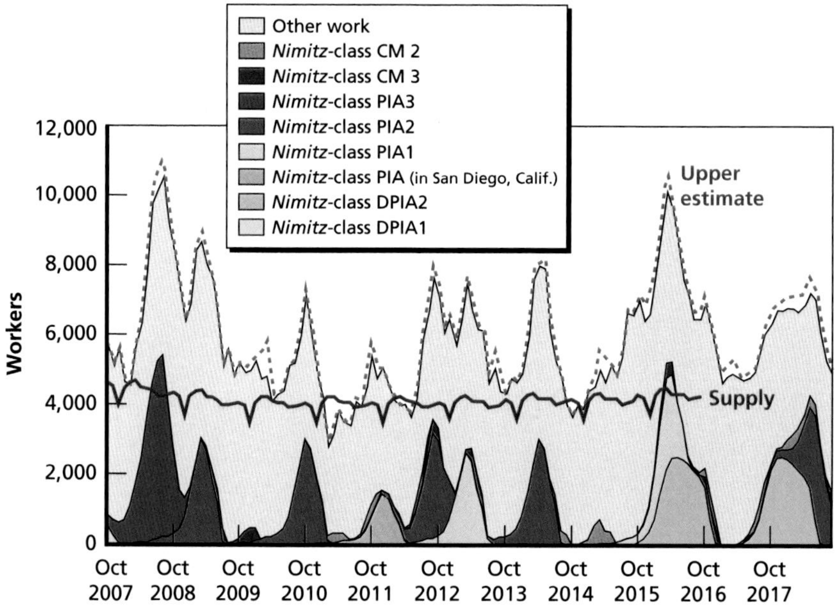

NOTE: Supply data provided by the WARR file used for NNSY did not extend past October 2016. CM = continuous maintenance.

RAND *MG706-A.6*

Figure A.7
Total Workload for 18/24-Month Carrier Cycle—PSNSY

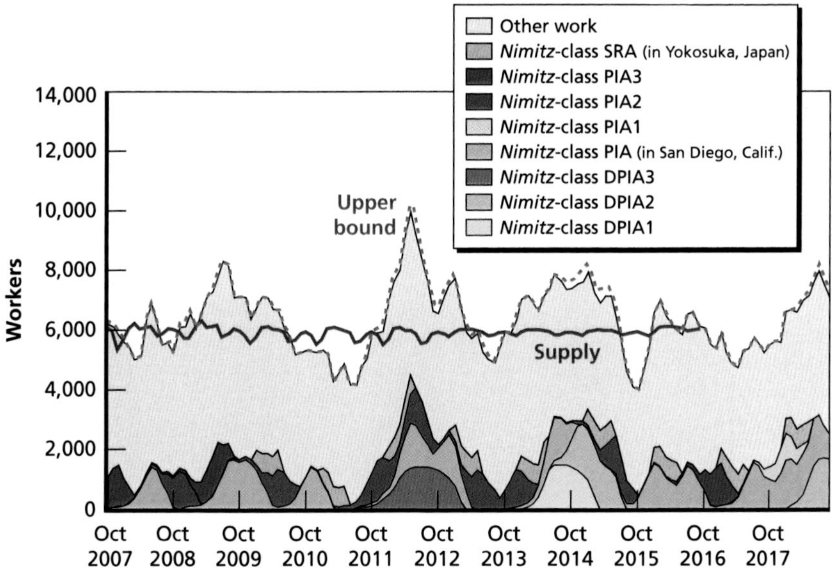

NOTE: Supply data provided by the WARR file used for PSNSY did not extend past October 2016.

RAND *MG706-A.7*

Figure A.8
Total Workload for 24-Month Carrier Cycle—PSNSY

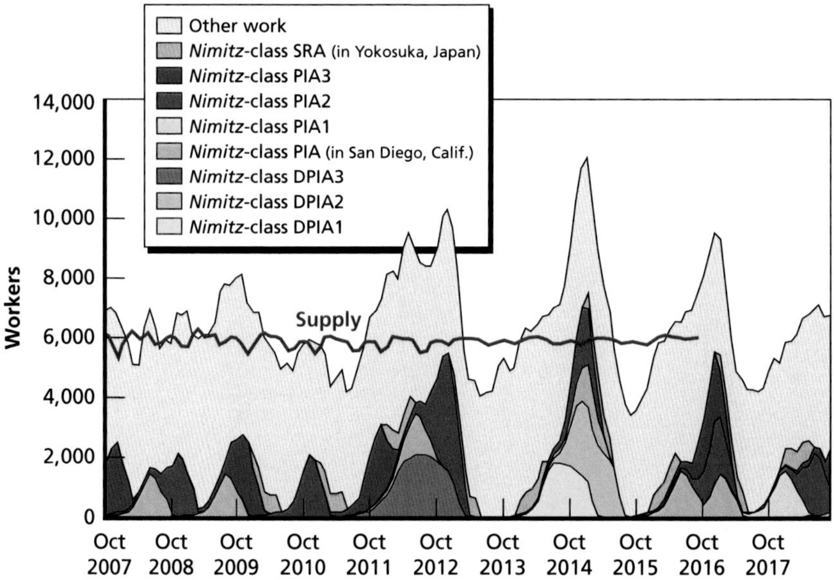

NOTE: Supply data provided by the WARR file used for PSNSY did not extend past October 2016.

RAND *MG706-A.8*

Figure A.9
Total Workload for 27-Month Carrier Cycle—PSNSY

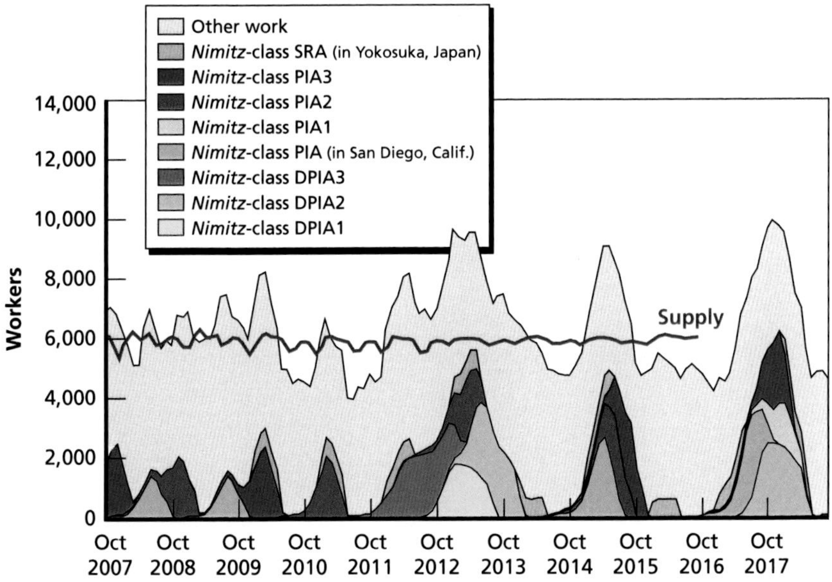

NOTE: Supply data provided by the WARR file used for PSNSY did not extend past October 2016.

RAND *MG706-A.9*

Figure A.10
Total Workload for 32-Month Carrier Cycle—PSNSY

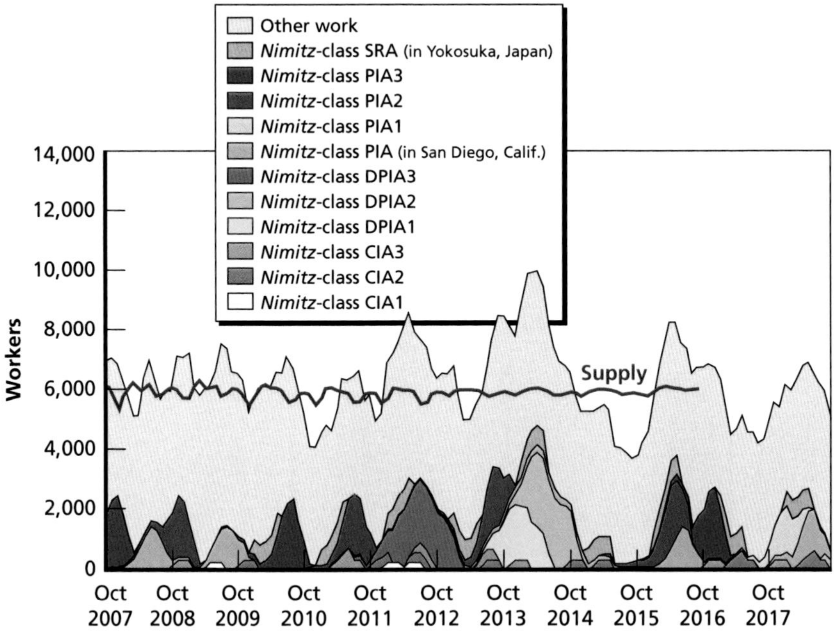

NOTE: Supply data provided by the WARR file used for PSNSY did not extend past October 2016.

RAND *MG706-A.10*

Figure A.11
Total Workload for 36/42-Month Carrier Cycle—PSNSY

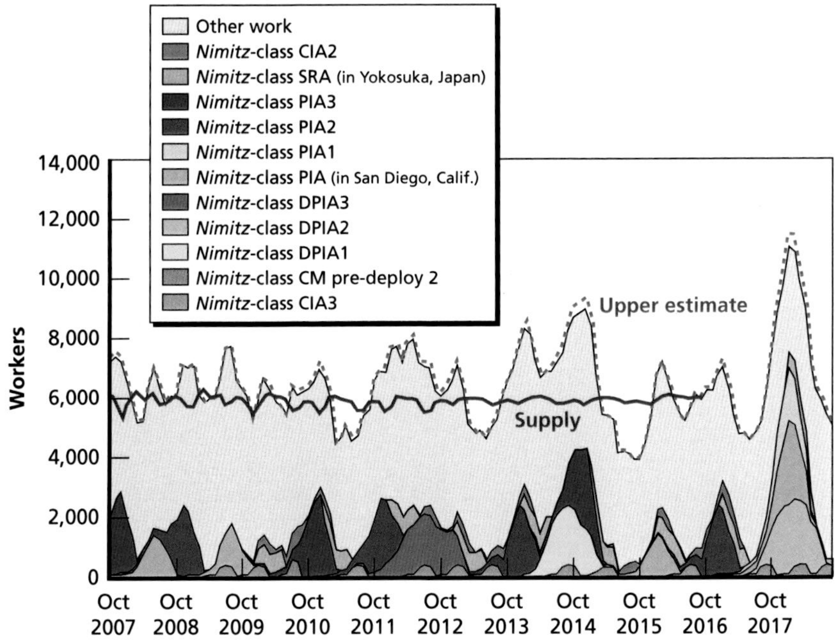

NOTE: Supply data provided by the WARR file used for PSNSY did not extend past October 2016.

RAND *MG706-A.11*

Figure A.12
Total Workload for 42-Month Carrier Cycle—PSNSY

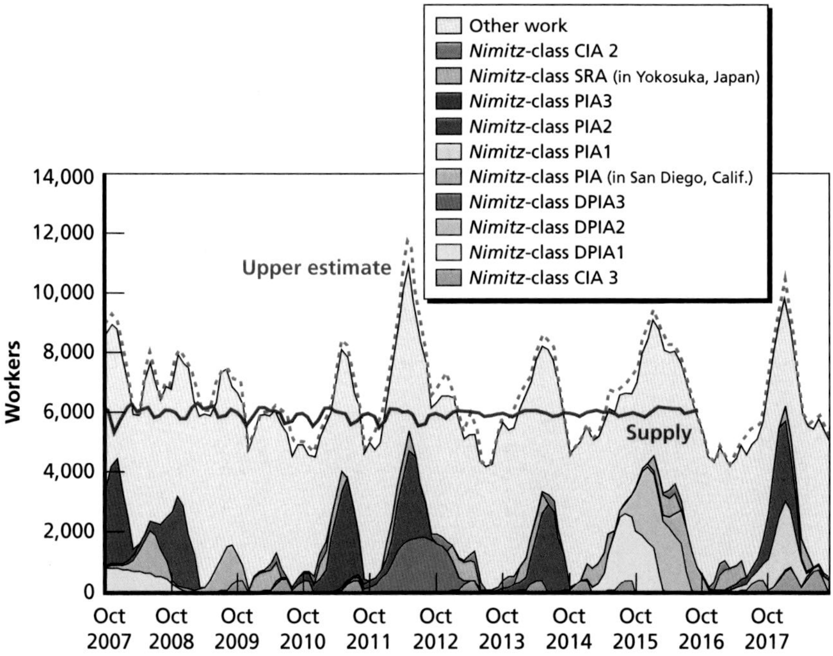

NOTE: Supply data provided by the WARR file used for PSNSY did not extend past October 2016.

RAND *MG706-A.12*

Bibliography

Arena, Mark V., John F. Schank, and Megan Abbott, *The Shipbuilding and Force Structure Analysis Tool: A User's Guide*, Santa Monica, Calif.: RAND Corporation, MR-1743-NAVY, 2004. As of January 8, 2008:
http://www.rand.org/pubs/monograph_reports/MR1743/

Behrens, Bruce H., Michael J. Bennett, Michael W. Price, and Delwyn L. Gilmore, *Long-Term Trends in Aircraft Carrier Employment,* Alexandria, Va.: Center for Naval Analyses, CRM D0003975.A2, October 2001.

Birkler, John, Michael Mattock, John F. Schank, Giles K. Smith, Fred Timson, James Chiesa, Bruce Woodyard, Malcolm MacKinnon, and Denis Rushworth, *The U.S. Aircraft Carrier Industrial Base: Force Structure, Cost, Schedule, and Technology Issues for CVN 77*, Santa Monica, Calif.: RAND Corporation, MR-948-NAVY/OSD, 1998. As of January 8, 2008:
http://www.rand.org/pubs/monograph_reports/MR948/

Carrier Planning Activity, "USS Nimitz (CVN68)-Class Aircraft Carrier 32-Month Maintenance Cycle Notional Analysis," unpublished analysis, Chesapeake, Va., April 23, 2006.

Congressional Budget Office, "Improving the Efficiency of Forward Presence by Aircraft Carriers," Washington, D.C., August 1996.

Department of the Navy, OPNAV Notice 3000.13B, N312, "Personnel Tempo of Operations," Washington, D.C., February 11, 2000.

———, OPNAV Notice 4700, "Representative Intervals, Durations, Maintenance Cycles, and Repair Mandays for Depot Level Maintenance Availabilities of U.S. Navy Ships," Washington, D.C., May 1, 2002.

———, COMNAVAIRFORINST 3500.20A, "Aircraft Carrier Training and Readiness Manual," San Diego, Calif., January 2005a.

———, OPNAV Notice 4700, "Representative Intervals, Durations, Maintenance Cycles, and Repair Mandays for Depot Level Maintenance Availabilities of U.S. Navy Ships," Washington, D.C., June 13, 2005b.

———, 4790, Ser 1800/344, "Review of Proposed USS NIMITZ (CVN 68) Availability Sequencing," Newport News, Va., November 17, 2005c.

———, Chief of Naval Operations Instruction 3000.15, "Fleet Response Plan," Washington, D.C., August 31, 2006a.

———, OPNAV Notice 4700, "Representative Intervals, Durations, Maintenance Cycles, and Repair Mandays for Depot Level Maintenance Availabilities of U.S. Navy Ships," Washington, D.C., August 31, 2006b.

———, OPNAV Notice 3000.13C, "Personnel Tempo of Operations Program," Washington, D.C., January 16, 2007a.

———, COMFLTFORCOMINST 4790.3, Revision A, Change 6, "Joint Fleet Maintenance Manual," Portsmouth, N.H., February 26, 2007b.

———, OPNAV Notice 4700, "Representative Intervals, Durations, Maintenance Cycles, and Repair Mandays for Depot Level Maintenance Availabilities of U.S. Navy Ships," Washington, D.C., August 31, 2007c.

Director, Warfare Integration, Office of the Chief of Naval Operations, "Report to Congress on Annual Long-Range Plan for Construction of Naval Vessels for FY 2007," Draft–Pre-decisional, December 30, 2005.

———, Report to Congress on Annual Long-Range Plan for Construction of Naval Vessels for FY 2008, February 2007. Dorsey, Jack, "Navy Changes Deployment Terms for First Time in 22 Years," *Norfolk Virginian-Pilot*, March 9, 2007.

"Earl Industries Awarded $165.2M Navy Contract," *MarineLink.com*, December 22, 2005. As of May 16, 2007:
http://www.marinelink.com/Story/EarlIndustriesAwarded%24165.2MNavyContract-201365.html

General Accounting Office, "Defense Logistics: GAO's Observations on Maintenance Aspects of the Navy's Fleet Response Plan," Washington, D.C., GAO-04-724R, June 18, 2004. As of January 31, 2008:
http://www.gao.gov/new.items/d04724r.pdf

Gomori, M. A., "USS Nimitz (CVN 68) Class Aircraft Carrier 32-Month Maintenance Cycle Notional Analysis," letter to Program Executive Officer, Aircraft Carriers, Newport News, Va., April 23, 2006.

Gordon IV, John, Peter A. Wilson, John Birkler, Steven Boraz, and Gordon T. Lee, *Leveraging America's Aircraft Carrier Capabilities: Exploring New Combat and Noncombat Roles and Missions for the U.S. Carrier Fleet*, Santa Monica, Calif.: RAND Corporation, MG-448-NAVY, 2006. As of January 8, 2008:
http://www.rand.org/pubs/monographs/MG448/

Hall, Matthew H., *The Impact of Long-Term Aircraft Carrier Maintenance Scheduling on the Fleet Readiness Plan*, thesis, Monterey, Calif.: Naval Postgraduate School, September 2004.

Harrington, Afi D., and Michael D. Bowes, *Submarine Depot Maintenance: Why Are Costs Increasing?* Alexandria, Va.: Center for Naval Analyses, CRM D0015095. A2, December 2006.

House Armed Services Committee, *Statement of Vice Admiral Justin D. McCarthy, SC, U.S. Navy, Before the Readiness Subcommittee of the House Armed Services Committee*, Washington, D.C., April 6, 2006.

Planning, Engineering, Repairs, Alterations—Aircraft Carriers (PERA-CV), *Incremental Maintenance Program for CVN-68 Class Aircraft Carriers*, Bremerton, Wash., May 3, 1996.

———, *Incremental Maintenance Program for CVN-68 Class Aircraft Carriers*, Bremerton, Wash., August 2002.

Naval Sea Systems Command, "Integrated Project Teams for Aircraft Carrier Maintenance," 3rd edition, Washington, D.C., March 2004.

———, "Aircraft Carrier Class Maintenance Plan," Washington, D.C., December 19, 2005.

Schank, John, Giles Smith, Brien Alkire, Mark V. Arena, John Birkler, James Chiesa, Edward Keating, and Lara Schmidt, *Modernizing the U.S. Aircraft Carrier Fleet: Accelerating CVN 21 Production Versus Mid-Life Refueling*, Santa Monica, Calif.: RAND Corporation, MG-289-NAVY, 2005. As of January 8, 2008: http://www.rand.org/pubs/monographs/MG289/

Senate Armed Services Committee, *Statement of Admiral Michael G. Mullen, Chief of Naval Operations, on the Fleet Response Plan*, Washington, D.C., March 9, 2006.

Sullivan, P.E., Rear Admiral, U.S. Navy, Washington Navy Yard, D.C., "Aircraft Carrier Drydocking Interval," letter to Program Executive Office, Aircraft Carriers, March 3, 2005.

Supervisor of Shipbuilding, Conversion and Repair, U.S. Navy, Newport News, Carrier Planning Activity (CPA), Chesapeake, Virginia, Code 1800, "USS NIMITZ (CVN 68) Class Aircraft Carrier 32-Month Maintenance Cycle Notional Analysis," April 23, 2006.

Work, Robert, "Know When to Hold 'Em: Modernizing the Navy's Surface Battle Line," Center for Strategic and Budgetary Assessments, Washington, D.C., September 20, 2006. As of January 29, 2008: http://www.csbaonline.org/4Publications/PubLibrary/B.20060920.Modernizing_the_Na/B.20060920.Modernizing_the_Na.pdf

Yardley, Roland J., John F. Schank, James G. Kallimani, Raj Raman, and Clifford A. Grammich, *A Methodology For Estimating the Effect of Aircraft Carrier*

Operational Cycles on the Maintenance Industrial Base, Santa Monica, Calif.: RAND Corporation, TR-480-NAVY, 2007. As of January 8, 2008: http://www.rand.org/pubs/technical_reports/TR480/